THE WORLD'S CLASSICS

WHAT MAISIE KNEW

HENRY JAMES was born in New York in 1843 of
ancestry originally both Irish and Scottish. He
received a remarkably cosmopolitan education in
New York, London, Paris, and Geneva, and entered
law school at Harvard in 1862. After 1866, he lived
mostly in Europe, at first writing critical articles,
reviews, and short stories for American periodicals.
He lived in London for more than twenty years, and
in 1898 moved to Rye, where his later novels were
written. Under the influence of an ardent sympathy
for the British cause in the First World War, Henry
James was in 1915 naturalized a British subject. He
died in 1916.

In his early novels, which include *Roderick Hudson*
(1875) and *The Portrait of a Lady* (1881), he was
chiefly concerned with the impact of the older
civilization of Europe upon American life. He
analysed English character with extreme subtlety in
such novels as *What Maisie Knew* (1897) and *The
Awkward Age* (1899). In his last three great novels,
The Wings of a Dove (1902), *The Ambassadors* (1903),
and *The Golden Bowl* (1904), he returned to the
'international' theme of the contrast of American
and European character.

DOUGLAS JEFFERSON is Professor of English Literature at Leeds University. DOUGLAS GRANT was
formerly editor of Classical American Texts (OUP).

THE WORLD'S CLASSICS

═══

HENRY JAMES
What Maisie Knew

═══

Edited by
DOUGLAS JEFFERSON AND DOUGLAS GRANT

Oxford New York
OXFORD UNIVERSITY PRESS

Oxford University Press, Walton Street, Oxford OX2 6DP

Oxford New York Toronto
Delhi Bombay Calcutta Madras Karachi
Petaling Jaya Singapore Hong Kong Tokyo
Nairobi Dar es Salaam Cape Town
Melbourne Auckland

and associated companies in
Berlin Ibadan

Oxford is a trade mark of Oxford University Press

New York edition of What Maisie Knew first published 1908
First published by Oxford University Press 1966
First issued as a World's Classics paperback 1980
Reprinted 1982, 1985, 1987, 1989, 1991, 1992

British Library Cataloguing in Publication Data

James, Henry
What Maisie knew.—(World's classics).
I. Title
II. Jefferson, Douglas William
813'.4   PS2116.W5   79-41271

ISBN 0-19-281533-4

Printed in Great Britain by
BPCC Hazells Ltd
Aylesbury, Bucks.

ACKNOWLEDGEMENTS

Acknowledgements are due to Messrs. Charles Scribners Sons for permission to use the New York edition of *What Maisie Knew* and the Preface by Henry James.

CONTENTS

CONTENTS

INTRODUCTION

TOWARDS the end of the eighteen-eighties James was approaching a period of painful crisis in his career as a novelist. Not that his reputation was declining, if we judge from the critical reception of his work, but the material rewards were diminishing. After *The Tragic Muse* (1889–90) he published no novels for six years; and although his short stories of this period include some of his finest he also wrote pieces like 'The Death of the Lion' and 'The Next Time', in which his sense of alienation from the great reading public finds a not altogether healthy expression. The recurring theme of the supersensitive aesthete suffering for his standards in a world of blatant vulgarity does justice to neither side. At no other time in his career did James allow his unresolved problems so to affect the emotional tone of his work. His chief preoccupation during these years was with the theatre, and his even greater failure here, artistic as well as worldly, intensified the crisis. The dreadful moment in January 1895 when he faced the hostile elements in the theatre on the first night of *Guy Domville* brought it to a head. His response to this shock, as the notebook entries and letters of the next few weeks reveal, was admirable.[1] He returned to the novel with renewed hope and purpose, and before the end of January was writing in the private, self-communing manner of the notebooks: 'I take up my *own* old pen again . . . large and full and high the future still opens.' In his desire to retrieve something from the wreckage of his theatrical ventures he stressed one lesson enforced by these years: 'the divine principle of the scenario'; his habit of working out a detailed and concrete plan of the sequence of events in a composition. The idea was not new: the first item in the notebooks (1878) is

[1] For the documentation and discussion of this phase students of James are much indebted to the editors of the *Notebooks* (1947), F. O. Matthiessen and Kenneth B. Murdock, and to Leon Edel for his indispensable Introduction to *The Plays of Henry James* (1949).

a fairly full sketch of the novel *Confidence*, where certain scenes are designed with a measure of detail. But the scenarios for *The Spoils of Poynton* and *What Maisie Knew* (both novels appeared in 1897) show a considerable intensification of the method, more wrestling with difficulties, and a more specific concern with the way his intentions could be realized in the ordering of the episodes. There is little resemblance between James's plays and his later fiction, but the effort to think as a dramatist was no doubt valuable to him as a novelist. The theatre imposes an especially rigorous economy: the author is sharply confronted with the question of which elements in his story can be rendered scenically.

Whether or not the technique of the 'scenario' is responsible for the difference, there is no doubt of the technical superiority of *What Maisie Knew* over, say, *The Tragic Muse*. The latter novel is quite well shaped by ordinary standards, but rather diffuse and uneconomical compared with James's later masterpieces. It is full of long conversations that too nearly resemble other long conversations, scenes in which the characters go once again over themes already fully discussed. In variety of invention and vivacity in the planning of episode with episode *Maisie* is in quite a different class.

The first idea for the story came to him in conversation at dinner with James Bryce and others in 1892. He states in the preface that, in its original form, the theme was simply that of a child of divorced parents 'entertained in rotation by its father and mother', and the unfortunate situation arising when one of them married again, providing it with an unsympathetic step-parent. Already, in the first notebook entry, James begins to experiment with new patterns: perhaps the parents could both re-marry and both die, the step-parents then marrying and looking after the child; or, better still, the parents might live, their indifference to the child bringing the step-parents together in a common responsibility but also as lovers. He takes the idea up again about nine months later (August 1893), developing the second possibility a little further but without much detail. As late as December 1895 he still thinks of it as material for a five-thousand-word story, though this becomes ten thousand in the longer note later in the same month. Here we have the first detailed scenario. Certain recognizable figures appear: the pretty governess who changes sides and marries the father; the 'frumpish' governess who finally rescues the child; the mother's second husband, the 'Captain' as he is designated here: 'simple, good, mild chap, bullied, hustled by his wife' and tenderly attached to Maisie. At this stage the 'centre of consciousness' is specified: the central subject will be derived from 'the dim, sweet, scared, wondering, clinging perception of the child'. The 'strange complicating action of the child's lovability' is now seen as 'the essence of the drama'. In October 1896 a new scenario appears in which we learn that the first eight chapters have been written, tallying with the first eight of the final composition, and that he planned to bring the novel to a close in four or five more chapters. As late as December 1896 a note appeared in the *Bookman* describing the forthcoming publication as 'a 25,000 word novelette'. The expansion to thirty-one chapters and something of the order of 90,000 words belongs to a relatively late stage of composition and the notebooks make no reference to it. What he had not foreseen was the much fuller development of the later incidents, with the growth of Maisie's awareness—for example, the scene with her father and his unsightly paramour takes up two chapters—and, more particularly, the splendid possibilities in the final section where Maisie, after much talk with Mrs. Wix and a last, lingering episode with Sir Claude, makes her choice. Originally Mrs. Wix was to have settled the matter by descending upon the illicit couple to 'save' Maisie, and this was to have occupied a chapter. In the novel the Boulogne episode—the setting was another happy afterthought—occupies ten chapters, hardly anything of which is hinted at in the notebook entries. It is evident that James's conception of Maisie took great strides forward during the final stages of composition.

Comparison with the scenarios reveals interesting improvements in the designing of episodes. James had planned that Maisie and her step-father should meet Ida in Kensington Gardens, with a 'strange gentleman', who, 'at Ida's request' walks away with the child but says nothing to her. And he had arranged a closely parallel scene in which Maisie and her step-mother meet her father and 'a strange lady', who also takes Maisie aside while the father puts on an act of 'insincere jealousy' similar to Ida's performance with her spouse in the earlier passage. In the novel itself much more is made of both incidents, and some of the elementary points of similarity and symmetry originally intended are avoided. Maisie's session with the Captain in the park becomes the occasion for new, unexpected beguilements and revelations. Stepping, as it were, through the looking-glass, she finds herself in a world where values are changed. Her mother, whom she has come to regard as indifferent to her and scarcely a part of her life, is now spoken of in words that wring her heart. The second episode, where Maisie is whisked away from the Earl's Court Exhibition to the fascinating apartment of the 'Countess', also has an atmosphere of beguilement but in the style of the *Arabian Nights*. To Maisie's imagination the two halves of the evening's outing are linked. She had come, with Mrs. Beale, in the hope of exotic entertainment, but as usual the sixpences had failed. Now 'everything she had missed at the side-shows was made up to her by the Countess's luxuries'. The Countess herself, awful as she later turns out to be, first comes into view near the garish booth of the 'Flowers of the Forest' (a bevy of 'bright, brown tropical ladies') and Maisie begins by taking her for one of these.

The two passages, at first intended as parallel to each other, are only partially so. The second, where Maisie's father contrives a leave-taking of extraordinary baseness, finds a partner and rival in the later one at Folkestone, in which her mother achieves the same end. Such elements of symmetry as these grouped episodes provide have value for more than decorative reasons. It is part of Maisie's fate that history keeps repeating itself. Having parents similar in habits and character, equal in their commitment to her and their desire to be rid of her, she lives in a world of recurring patterns which serve her as opportunities to learn from experience. When the patterns recur it is with a difference, for Maisie grows in perception. With the Captain in the park we see her as still a very young child, completely under the spell of her encounter and with a heart that opens without reserve to the appeal of kind words. Between this episode and her scene with Beale she has developed somewhat. Still as responsive to wonderful appearances—her father's presence dazzles her—she now has the penetration to see at least something of what his awkwardness signifies, as she sits on his knee and he strokes her hair, uttering pointless endearments and trying to find something to talk about. This is a truly formidable passage:

There was something in him that seemed, and quite touchingly, to ask her to help him to pretend—pretend he knew enough about her life and her education, her means of subsistence and her view of himself, to give the questions he could n't put her a natural domestic tone. She would have pretended with ecstasy if he could only have given her the cue. She waited for it while, between his big teeth, he breathed the sighs she did n't know to be stupid. And as if, though he was so stupid all through, he had let the friendly suffusion of her eyes yet tell him she was ready for anything, he floundered about, wondering what the devil he could lay hold of.

In her final encounter with her mother she is still child enough to make the endearing error of recalling for her benefit the words of the Captain; but her tender wish to say something relevant to her mother's point of view proceeds partly from a recognition of the latter's need to be humoured. Ida's incoherent and hopelessly self-centred monologue, which this friendly remark punctures, is addressed to an older Maisie, who sometimes has difficulty in keeping the look of wonder out of her eyes. With her glimpses of what Ida's repudiation of the Captain (now 'the biggest cad in London') might mean, Maisie's sympathetic understanding is stretched as never before.

The fact that all the adults, in their turn, absent themselves from Maisie's life for a time, and then find her more of a person to be reckoned with on renewal of acquaintance, provides James with a happy means of registering her growth. These reunions also give opportunities for the kind of irony in which this novel abounds. The adults are far from indifferent to the impression they make on Maisie. They all want to be loved by her, and Maisie's love is there to be tapped by all who wish for it. When she returns to Mrs. Beale after an unusually long absence, that beauty, more florid than ever to Maisie's vision (the child's perception has grown, and Mrs. Beale has also changed somewhat), 'fairly swooped upon her' and brought it home to her 'how quite formidably, indeed, after all, she was loved':

"You're too delicious, my own pet [she says in response to one of Maisie's unintentional witticisms]. . . . How in the world have I got on so long without you? I've not been happy, love. . . ."
"Be happy now!"—Maisie throbbed with shy tenderness.

'My own child', her mother unexpectedly murmurs, in 'a voice of sudden confused tenderness', when Maisie and Sir Claude discover her with the Captain; and the child is thrilled by the first maternal appeal she has ever experienced. 'The next moment she was on her mother's breast, where, amid a wilderness of trinkets, she felt as if she had suddenly been thrust, with a smash of glass, into a jeweller's shop-front', only to be as suddenly pushed into the arms of the Captain, for whom this demonstration is largely intended. And when, at Folkestone, Maisie suggests that her mother (who is about to break with her for ever) should stay for tea, she receives a kiss on the brow. 'She *is* sweet!' Ida says to Sir Claude.

Maisie's increased grasp of situations, and the greater scope of her relationships with the adults, all of which is accommodated by lengthier treatment of scenes and episodes, point to a growth which is not precisely measured for us in time. So long as the six-monthly arrangement is adhered to we can count her years; but there is an unspecified period during which Ida neglects to take her, followed by one in which Beale shows an equal indifference to his commitment; and to these vaguenesses others are added. During the greater part of the novel we do not quite know her age, and there is no reason why we should. James, in fact, takes liberties. The time-table of the later events, examined strictly, allows a lapse of only a few weeks between the episode in Kensington Gardens and the final scene; but the growth in Maisie's perceptions during this period would suggest a much longer stretch. References to time are so unobtrusive that this illusion of a rather older Maisie does not appear to be contradicted. The lack of precision has advantages. Maisie has advanced in some ways and not perhaps in others. She is a very special child. 'It was the fate of this patient little girl . . . even at first to understand much more than any little girl, however patient, had perhaps ever understood before.'

Whatever she has come to 'know' by the end of the book, it does not include the kind of knowledge more appropriate to puberty. This can be quite simply asserted, not because the novel tells us but because its logic demands it. If Maisie really knew more all her perspectives would change and the book would not have the unity we find in it. What Maisie actually thinks about intimate relationships between adults remains her secret. (She has a phrase—'lady-intimate'—for Mrs. Beale in her relation to Sir Claude.) There is no evidence that she sees any harm in them, after all Mrs. Wix's probings and pouncings: ' "*Is* it a crime?" Maisie then asked. Mrs. Wix was as prompt as if she had been crouching in a lair. "Branded by the Bible." ' Even after this pronouncement, and others equally lurid, she can still hanker after more charming interpretations of the situation, which provoke Mrs. Wix to something verging on panic. Some critics have deplored Mrs. Wix's limited conventional outlook and its effect on Maisie, but this seems unnecessary. They are, each of them, as good as gold, and no harm will be done by their misunderstandings, of which James makes fine comedy. F. R. Leavis says rightly that, after such a childhood, it is well that Maisie should enter adolescence 'under that kind of respectable tutelage'.

If there remains much that Maisie doesn't know, and

much that we never learn about her reasons, her correct decision in the last chapter is nevertheless intelligible enough. She realizes, on Sir Claude's return to Boulogne, that all is not well, not only because of Mrs. Wix's distress but because Sir Claude greets her so uncertainly and obviously lies to her about his not having been with Mrs. Beale; and when they go off together for breakfast, as if their old easy relationship were renewed, she sees that behind the make-believe lies a fear: '. . . his fear was sweet to her, beautiful and tender to her, was having coffee and buttered rolls and talk and laughter that were no talk and laughter at all with her; his fear was in his jesting postponing perverting voice. . . .' And when Sir Claude tries hopelessly to outline to her the little programme that the three of them—Mrs. Beale, himself, and Maisie—might work out together ('"Of course it would be quite unconventional . . . but things have got beyond that, don't you see?"'), doing justice in turn to Mrs. Wix's scruples, the 'good' that she and Maisie have done him, and the fact that Mrs. Beale and he will have to look after her education themselves, since no governess would stay ("of the only kind that would do"): when Sir Claude, with his 'mottled face and embarrassed but supplicating eyes', puts all this to her, how can she not know that something is wrong?

Our limited knowledge of what Maisie 'knew' brings up the familiar technical issue of James's use of her as 'centre of consciousness'. An intenser application of the 'centre of consciousness' method is one of the significant aspects of his changing technique at this period. He had been using it, of course, from the beginning of his career, but in these later works the imaginative experience of the central character is treated with greater inwardness. External narrative becomes more intimately adjusted to the tone of the mental adventure, and all that happens is saturated with the life of the mind. But another component, often ignored, must be taken into account: the interpretative contribution of the narrator himself, whose imaginative involvement in the events and in the process of narration sometimes operates in close association with the central character's point of view but often at some distance from it. The reader's attention, in fact, is a good deal less constantly and less completely focused on the consciousness of Maisie, Strether, Maggie, and the others than we are often led to suppose. Much of the art of the later books lies in a kind of equivocation. The inwardness of certain passages creates the illusion of a pervasive use of the single point of view, and this is sufficiently well sustained to allow the narrator a good deal of freedom. The novelist's personal wit and rhetoric can be remarkably active without any serious disturbance of this impression.

The mental gap between adult narrator and child heroine makes this novel a special case, and in the preface to *What Maisie Knew* the role of the narrator is referred to more than in the other prefaces; but even so James does not do it the fullest justice. His emphasis—and who would wish to blame him?—is on the role of Maisie, the effect of her 'freshness', her capacity for wonder, on the presentation of the persons and events she witnesses. On this theme James expresses himself with the greatest tenderness and eloquence: nowhere has an artist written more movingly of his own achievement. What he does not show is that this result has been brought about with extraordinarily little exposure of Maisie's inner being. He writes that 'it is her relation, her activity of spirit, that determines all our concern—we simply take advantage of these things better than she herself', which is an important part of the truth. But he does not indicate how far he allows her direct sense of things to lapse. Present throughout the novel as narrator, but in Maisie's interest, he makes the adult tone work on her behalf. Selected moments or aspects of her awareness are indeed used with magical effect, and the charm lingers during passages when the child's mind is closed to us. Most of what may be called her private life remains her own affair. Her words often have the freshness of a revelation, coming from processes of thought to which we have had no access.

The contribution that Maisie's special viewpoint makes is nowhere more remarkable than in her image of her

mother—'concrete, immense and awful', as James describes
her in the preface. The child's imagination heightens the
physically portentous—her 'violent splendour, the wonderful colour of her lips . . . the hard stare, the stare of some
gorgeous idol described in a story-book'—while reducing
what is morally shocking to the merely bizarre. But her view
of some situations is constantly supplemented by the interpretations of adults, notably Mrs. Wix and Sir Claude, who
tell her many things that she must appropriate at her own
level of understanding and of which she sometimes makes
strange capital. James saw the use that could be made of
Mrs. Wix when he wrote his final notebook entry for *Maisie*:
'Don't I see the whole thing reflected in the talk, the confidences, the intercourse of Mrs. Wix? Something very
pretty may be made of this—her going a little further and
further, in the way of communication, of "crudity", with
the child than her own old-fashioned conscience quite war-
rants—her helpless sighs at what she has perforce to tell her,
at what Maisie already has seen and learnt—so that she
doesn't make her any more initiated, any "worse" etc., etc.,
and thus serves as a sort of dim, crooked little reflector of the
conditions that I desire to present on the part of others.' But
Mrs. Wix is also the source of some of her most delightful
notions. She learns from her that Sir Claude is 'more of a
gentleman than anybody else in the world'. Stimulated by
fond affection, Mrs. Wix's imagination expatiates on his
possibilities and his future: 'He's a wonderful nature, but he
can't live like the lilies . . . he must have a high interest.' In
fact they—Maisie and she—must get him into Parliament.
Her idea that Sir Claude needs to be 'saved' comes at first
as a startling one to Maisie, but it captivates her when Mrs.
Wix associates it with an appeal to him to devote himself to
Maisie. Not realizing that Mrs. Beale represents the menace
from which he must be saved, Maisie recalls Mrs. Wix's pronouncement when her stepmother is present. The latter also
has the idea that Maisie saves her, but in the other sense of
providing a pretext for illicit meetings with Sir Claude. The
whole dialectic of the child's relationship with the step-

parents is expressed in the double meanings of the following admirable passage:

> "As I'm saving Sir Claude?" the little girl asked eagerly. Mrs. Beale, a trifle at a loss, appealed to her visitor, "Is she really?"
>
> He showed high amusement at Maisie's question. "It's dear Mrs. Wix's idea. There may be something in it."
>
> "He makes me his duty—he makes me his life," Maisie set forth to her stepmother.
>
> "Why that's what I want to do!"—Mrs. Beale, so anticipated, turned pink with astonishment.
>
> "Well, you can do it together. Then he'll have to come!"
>
> Mrs. Beale by this time had her young friend fairly in her lap and she smiled up at Sir Claude.
>
> "Shall we do it together?"
>
> His laughter had dropped, and for a moment he turned his handsome serious face not to his hostess but to his stepdaughter. "Well, it's rather more decent than some things. Upon my soul, the way things are going, it seems to me the only decency!" He had the air of arguing it out to Maisie, of presenting it, through an impulse of conscience, as a connexion in which they could honourably see her participate; though his plea of mere "decency" might well have appeared to fall below her rosy little vision. "If we're not good for *you*," he exclaimed, "I'll be hanged if I know who we shall be good for!"
>
> Mrs. Beale showed the child an intenser light. "I daresay you *will* save us—from one thing and another."

Maisie has the capacity to extract her own kind of satisfaction from what she learns. At an early stage the idea has been put into her head that she brings people together, and she cheerfully echoes the happy truth when Sir Claude says this in reference to himself and Mrs. Beale, upon which the latter recalls that she did it also for Beale and herself. Maisie reverts again and again to this achievement; and even at the end, when Sir Claude's inability to give up Mrs. Beale is the cause of the trouble, and Mrs. Beale's resentment at the turn of events has burst out in inexcusable violence, the latter has only to introduce the old cherished fact and Maisie echoes it, with the sweetest, friendliest irrelevance.

Sir Claude's habit of taking her into his confidence furnishes her with a number of ideas with which she must do what she can. He tells her that her mother has been 'squared', that she lets him do what he likes provided that he gives her the same freedom; and Maisie is pleased, although a little at a loss concerning the implications; and later she is ready with the same formula for Mrs. Beale in her relations with Beale. Maisie thinks it strange, and then makes it a matter for tenderness, that Sir Claude should be so afraid of women. He confesses without reserve—indeed with relish—to his fear of his wife; and, later, to his fear of Mrs. Beale. Living constantly in a world of imperfectly under-stood notions, Maisie develops an innocent tendency to claim more knowledge than she has; and this fills the good Mrs. Wix with the greatest alarm concerning her 'moral sense'.

James's characterization reaches its highest excellence in this novel. Sir Claude, seen through Maisie's adoring though observant eyes, is his finest study of the English upper-class man of leisure: an immense advance on the Lord Lambeths and Lord Warburtons of earlier works, good though these are. The difference is in the wealth of sensitive, humorous, touching detail. His inadequacies are handled very lightly. (His gift of a 'huge frosted cake' during a period of school-room shortage is noted as a matter for gratitude, the blame for the nonpayment of Mrs. Wix falling entirely upon Ida.) But nothing, not all his kindness and tact, can disguise the fact that he fails. He had sworn never to forsake Maisie, and this in effect is what he does at the end. He has about him a quite exceptional aura of niceness, with all his frailties; but the frailties are never lost sight of, and they are decisive.

Mrs. Wix has been described as 'Dickensian'. In the earlier chapters where we see her, through Maisie's very immature gaze, in terms of the 'straighteners', the diadem, the old brown dress, and her memories of Clara Matilda—but also her genuinely loving embraces—we may perhaps be reminded of the minor aspects of Dickens's art. Dickens's more inspired creations are a different matter, outside James's range. But Mrs. Wix develops into something that Dickens could not have touched. As the portrait gains in depth and reality, with her successive manifestations of steadfastness and aggressiveness on behalf of Maisie, her idiosyncrasies also become more remarkable. The curious passage where she probes Maisie's attitude to Sir Claude and Mrs. Beale contains an unexpectedly fine yet extravagant moment which illustrates what James could do with a character so far from his habitual world (or worlds) of manners. Maisie is responding to her promptings:

"If I thought she was unkind to him—I don't know *what* I should do!"

Mrs. Wix dropped one of her squints; she even confirmed it by a wild grunt. "I know what I should!"

Maisie at this felt that she lagged. "Well, I can think of *one* thing."

Maisie met her expression as if it were a game with forfeits for winking. "I'd *kill* her!" That at least, she hoped as she looked away, would guarantee her moral sense. She looked away, but her companion said nothing for so long that she at last turned her head again. Then she saw the straighteners all blurred with tears which after a little seemed to have sprung from her own eyes. There were tears in fact on both sides of the spectacles, and they were even so thick that it was presently all Maisie could do to make out through them that slowly, finally Mrs. Wix put forth a hand.

When Mrs. Wix finds speech it is to repeat 'I adore him! I adore him!', with which sentiment Maisie concurs.

No writer knew better than James how to convey the pleasure of being abroad for the first time, and it was a happy stroke to add this awakening to Maisie's other initiations. Her delighted response to Boulogne reminds us of James's own early impressions of France in 1855, recorded in *Notes of a Son and a Brother*. When we are told that 'she recognised, she understood, she adored and took possession; feeling herself attuned to everything and laying her hand, right and left, on what had simply been waiting for her', we are reminded also of James's 'passionate pilgrims', who come to Europe with an image already formed, fatally equipped for the real thing. Maisie's capacity for appreciation of 'abroad' greatly enhances the later chapters, especially those in which she has her last jaunt with Sir Claude. Aware though she now is of things that are bringing her childhood to a close, she still has the child's pleasure in an outing. She is fascinated by the café floors sprinkled with bran, like a circus, and by the picturesque old gentlemen who pick their teeth or soak their buttered rolls; and all the problems of their situation cannot destroy for her the charm of having breakfast in the French style with Sir Claude:

"Yes, I may as well confess to you that as much as that I do know. *She* won't go away. She'll stay."

"She'll stay. She'll stay," Maisie repeated.

"Just so. Won't you have some more coffee?"

"Yes, please."

"And another buttered roll?"

"Yes, please."

He signed to the hovering waiter, who arrived with the shining spout of plenty in either hand and with the friendliest interest in mademoiselle. "*Les tartines sont là.*" Their cups were replenished and, while he watched almost musingly the bubbles in the fragrant mixture, "Just so—just so," Sir Claude said again and again. "It's awfully awkward!" he exclaimed when the waiter had gone.

The 'homeliest notes' of travel never lost their magic for James. In *The Middle Years*, among his last writings, he could still recall his early delight at the way, in England, the plate of buttered muffin and its cover are 'sacredly set upon the slop-bowl' after hot water has been poured into the latter. That Maisie can make the most of her breakfast is characteristic of the way she takes so much of what happens to her. She is as healthy and resilient as she is loving and good, and for this reason the novel, for all its sombre and pathetic possibilities, has a wonderful lightness. Not that Maisie on this crucial occasion succeeds in maintaining the mood of enjoyment. When the great issue arises, and the pair are faced with a wretched parting, there is no question of any pleasant little *déjeuner* together.

James said on many occasions that it took an American to appreciate Europe to the full. *What Maisie Knew* is one of his 'English' novels, if we arrange his works according to subject-matter, but no English novelist could have cherished a figure like Sir Claude with James's intimate and poetic sense of type. The London settings are lightly sketched, but they are by the author of *English Hours*; and, as we have seen, Maisie's Boulogne is the richer for its affinities with the tradition of American travel. It is a peculiar and important characteristic of American literature that it achieves some of its greatest triumphs by taking over the material of another civilization and giving it at once a new heightening and a new subtlety. Since 1883 James had lived without interruption away from the United States and had tended more and more to choose English subjects for his fiction. 1888, the year that produced *The Reverberator*, *A London Life*, *The Aspern Papers*, and *The Patagonia*, all works with leading American characters, seems to mark the end of a phase, and during the next dozen years or so all his novels and most of his stories deal with English materials. (But it is worth noting that 'The Pupil' and 'Europe', the stories of this period which have American subjects, are especially fine.) With *The Wings of the Dove* (1902) the American theme returns with great splendour and dominates James's culminating phase as a novelist.

The English works are at their best perhaps when the situations are illuminated by an ethical awareness borrowed from James's American heritage. It has been noted before that Fleda Vetch in *The Spoils of Poynton* might have been a New England girl; while Maisie, with her acute, optimistic, humorous sense of life, surely has more than a touch of American childhood in her.

In recent years *What Maisie Knew* has come to be one of James's most admired novels, being singled out for high praise by critics who are otherwise not convinced by the claims made for his later development. For many years it suffered a measure of neglect, as did most of his work: the sheer quantity of James's output, and the special interest excited by a few favoured masterpieces, caused it to be over-shadowed. Lest we imagine that our appreciation today is so very much keener than that of his first readers we do well to turn back to some of the comments of the reviewers, making due allowance for the fact that the novel stands at the threshold of the late period and was likely to offer some difficulties even to the well-disposed. The subject-matter was an obstacle to many. The unruffled ease with which James presents the routines and subterfuges of the adulterous as part of the familiar conditions of life for a young child was bound to be widely misunderstood. But there were some highly favourable reviews. The *Bookman* praised 'the wonderful dramatic analysis of this marvellous book', and responded well to its tone: 'Mr. James keeps his sense of humour for he knows his little heroine, intelligent, uncorrupted, valiant, eager for life, will come through with an unbroken and a gentle spirit.' The *Pall Mall Gazette* reviewer said that, with this novel, James 'without doubt' touched 'his highest point', though he found the technique somewhat perplexing. The *Athenaeum* reviewer was divided between admiration for the novel and dislike of the theme. Maisie's constant involvement with her horrid elders is described as 'oppressive and painful', but the importance of the book 'must be certainly apparent to those on whom analysis of the first quality and delicate delineation are not thrown away'. The heroine emerges from her ordeal redeemed 'by the force of a singularly buoyant and innate grace of nature. . . . She retains a child's heart and mind at their sweetest.' The *Speaker* was disturbed by an over-sophistication of treatment but emphasized that here was a 'great novel' by a 'great artist'.

Of the American periodicals *Book News* found it 'one of the most remarkable novels in English for years'. The *Critic* referred to the 'extraordinary feat of transmuting the waste material of society into something hard and clean and brilliant', but echoed the divided attitude of many readers in the remark that 'the skill and tenderness with which Mr.' James has handled this unheard of plot go far toward winning his pardon for the atrocity of having devised it'. Some critics, of course, found it boring and excessively complicated. As an example of hostile criticism at its most stupid the *Literary World*'s notice deserves to be quoted: 'Its author exhibits not one ray of pity or dismay at this spectacle of a child with the pure current of its life thus poisoned at the source . . . every manly feeling, every possibility of generous sympathy, every comprehension of the higher standards, has become atrophied in Mr. James's nature.'

On the whole this was not too bad a start; but for many years criticism of *Maisie* is disappointing. The survey that follows can only be very selective. Joseph Warren Beach's study of James (1918), the first serious treatment of his work, falls short in its assessment of this novel. Beach says that 'the story of Maisie is crowded with appeal the most simply human', that 'the irony and pathos' of her history is treated with 'a fidelity to life and a vividness of realisation possible only to high poetic genius'; and yet he finds the novel not among the most important 'humanly' of James's works, and he places it in a category labelled 'technical exercises'. Why? Beach's difficulty is with the limitations of the point of view. We do not learn enough, he complains, about the adult side of the picture, and he cannot be satisfied with the child's viewpoint, charming though it is. There is something a little phlegmatic in Beach's not responding to all that James made of Maisie's viewpoint, and his treatment of the novel is by no means an advance on the comments of the earliest admirers.

Maisie does not appear among the chosen texts for Lubbock's *Craft of Fiction* nor for Forster's *Aspects of the Novel*, the best-known general works on fiction of the nineteen-twenties. Pelham Edgar's few pages of discussion mark no significant development. His most striking contribution is an odd comment on the ending: 'His [Sir Claude's] adhesion in the end to Mrs. Beale, involving as it does the sacrifice of Maisie, looks like the weakness of excessive amiability, but I think we are intended to interpret as an act of strength his encouragement to Maisie to depart with Mrs. Wix . . . one closes the book with the inevitable feeling that a more radiant Sir Claude, freed from all the agitations his weakness has engendered, will one day look in upon Mrs. Wix and Maisie in some quiet English retreat and remain.' In this period, and for some time after, interesting criticism on *Maisie* is hardly to be found. Matthiessen does not discuss it, and Dupee in his 'American Men of Letters' study of James is not at his best on this novel, which he describes as 'witty and remorseless', doubting whether it is 'not more of a torment than a pleasure for the reader'. Mrs. Wix he finds 'almost diabolically righteous'.

F. R. Leavis's admirable pages in response to Marius Bewley's views are a landmark in critical recognition. Bewley's case first calls for brief comment. We become aware in his work of one of the chief peculiarities of contemporary criticism, especially in America: that is, a considerable moral seriousness, with the object of moral concern misplaced. Bewley is a little worried by Mrs. Wix—for example, by her declaration that she 'adores' Sir Claude. Is she playing the same game, he wonders, as Mrs. Beale: that is, of using Maisie 'as a means of closing in' on that object of erotic desire? As Leavis says in his reply, '"erotic" in these days is a term of extensive and uncertain application' and surely 'a very odd term to apply to poor Mrs. Wix's state'. Bewley makes much of what he sees as the undesirable conventionalism of Mrs. Wix's morality, compared with Sir Claude's fine sensibility (as exemplified in his speech in the final chapter: 'it's the most beautiful thing I've ever met—it's exquisite, it's sacred'). And he predicts that Maisie, as Mrs. Wix's moral superior, will educate her governess. But surely Sir Claude's speech is that of a man who has given up the struggle and taken refuge in aesthetic rapture. James makes this clear: 'He was rapidly recovering himself on this basis of fine appreciation.' To insist in this context on Mrs. Wix's limitations (she is, after all, taking the responsibility for Maisie) and to refine upon the moral development of Maisie at her expense, are two pieces of characteristic modern preciosity. Another kind of preciosity, common in certain schools of criticism, is the undue heightening (or darkening) of moral significances, the uncalled-for use of words like 'evil'. Leavis again provides the sensible reply to Bewley in his refusal to see 'moral horror' in the novel. 'Moral squalor' is, for him, a strong enough description of the milieu in which Maisie lives, and he emphasizes that the book has the tone of comedy. But with all its aberrations Bewley's discussion is serious, and *What Maisie Knew* emerges from it as a great novel. Leavis's treatment, perhaps his best piece of Jamesian criticism, is a model of rightness and proportion.

William Walsh's excellent discussion of *Maisie* in *The Use of Imagination*, a book 'addressed to two classes of readers, those interested in literature and those interested in education', deals especially with the growth of the child's imaginative and moral insight. As a study of the psychological content it goes further than any other contribution, and it contains some fine critical comment. Of all its critics Leavis and Walsh have done most to establish the true interest and value of this novel. To these should be added Peter Coveney, whose interesting study of *What Maisie Knew* in *Poor Monkey* (1957) appears again in the revised edition of that book entitled *The Image of Childhood*, Penguin, 1967.

Others in recent years have achieved interpretations of unprecedented absurdity. According to one view (let its exponents be unnamed) Maisie is in fact sexually corrupted, and when in the last chapter she says she will wait for Sir Claude 'on that old bench where you see the gold Virgin', the chosen rendezvous has symbolic importance and tells us what she will be prepared to offer. Apart from the grossness of the misreading, is it conceivable that James could have written, on the re-perusal of such a story, that his little heroine really keeps 'the torch of virtue alive in an air tending infinitely to smother it', and that 'through the mere fact of presence' she sows 'the seeds of the moral life'? The best readings of the novel, as one might expect, are those that are most akin to James's great preface.

DOUGLAS JEFFERSON

NOTE ON THE TEXT

What Maisie Knew appeared in two serial versions: in the Chicago *Chap Book*, 15 January–1 August 1897, and in the London *New Review*, February–September 1897. The *New Review* version bears the marks of slight revision, and it also suffered drastic abridgement from Chapter 18 onwards. There had been an early misunderstanding between James and the editor, W. E. Henley, about the total length. The latter's attitude to the whole undertaking is known to have been grudging, and the *New Review* was in financial difficulties.

Ward S. Worden, who has made a careful study of the abridgement, argues that James himself carried it out, and this seems sensible; but his claim that the abridgement is of some interest in its own right, as offering a different interpretation, must surely be rejected. Obviously the cuts make a difference to our view of Maisie, but it seems clear that James, for whom the operation must have been wholly distasteful, was primarily concerned with the clarity and coherence of the main narrative, like the practical craftsman he was. What he sacrificed, however interesting and excellent, was what could best be spared in the light of that consideration. Many sequences of dialogue are partially or entirely shorn of their accompanying passages of imaginative psychology, and this reduces them to a disconcerting baldness. In the scene between Maisie and her father the essential element of his equivocal offer is retained, but the last pages of Chapter 18, in which Beale is exposed so ruthlessly in his relationship with his daughter, are cut. All passages relating to the money Ida was apparently intending to give to Maisie during their last encounter are removed. The largest excision consists of two whole chapters, 27 and 28, which deal with Mrs. Beale's stay in Boulogne before Sir Claude's reappearance.

The first English edition in book form, published by William Heinemann, London, appeared in September 1897; and the first American edition, published by Herbert Stone & Co., Chicago and New York, in October 1897. They represent separate revisions of the serial versions, the changes (styles of punctuation apart) being at the rate roughly of one in every two or three pages, and confined to small matters of taste in vocabulary or word order. Where the two first editions differ the readings of the English are later, either agreeing with the New York text or replaced there by further revisions. In the course of his abridgement of the later chapters of the *New Review* version, James introduced a few readings which tally with or move in the direction of the first English edition, and so are later than the corresponding readings in the first American edition; but otherwise the *New Review* text is earlier than the latter. The differences between the two first editions are not much more numerous than the revisions of the serial versions, and they are of similar significance. One example of James's concern for the *mot juste* may be singled out. In the passage (Chapter 9) where Maisie's mother tries to wring information out of her and, according to her system, the child feigns stupidity, the *Chap Book* and *New Review* readings have: 'She had achieved a vagueness beyond her years.' In the first American edition 'vagueness' becomes 'vacancy', but in the first English edition 'hollowness' is substituted, and this remains in the New York text. The gentleman accompanying Ida when she calls on Mrs. Wix (Chapter 23) was named Mr. Knackfuss in the *Chap Book*, but he becomes Mr. Tischbein in subsequent texts. A minor curiosity, in Chapter 15, is the substitution, in the first American edition only, of the 'Baron' for the 'Count' in all other versions, earlier and later.

What Maisie Knew appeared in Volume XI (1908) of the New York edition of the Novels and Tales (published by Charles Scribner's Sons), along with *In the Cage* and 'The Pupil'. James revised it with the first English edition as his point of departure, and introduced quite a large number of changes: some two or three to the page. Some are very slight, others involve longish phrases, but very few embody significant modifications of meaning; and the latter are mainly local and hardly of major importance. On p. 28 'this sense of a tenderness', &c., has become 'this sense of a support, like a breast-high banister in a place of "drops"', &c., Mrs. Wix at this early stage in the novel being seen, in the revised text, as dependable rather than tender. James characteristically adds a novel image. On p. 49 the description of Sir Claude as 'the most radiant person with whom she had yet been concerned' has been changed to 'the most shining presence that had ever made her gape'. 'Radiant' is perhaps too suggestive of moral health, 'shining' being more compatible with unreliability. On p. 120 the Captain's final word has been changed from "Oh, I'll keep it up!" to "Oh I'm in for it". On p. 135 'his [Beale's] unexpected gentleness'—which does him too much honour—has been changed to 'the way he spared them'. There are no changes more far-reaching than these.

The list of variants on p. 269 includes a few more examples of changes of meaning, and also some places in which James has added a fanciful image or colloquial turn, or replaced abstract words by more concrete expressions. Not all of the changes can be attributed to any clearly marked tendency in his later style.

The present text is that of the New York edition. A few obvious misprints have been corrected.

A SHORT GUIDE TO
FURTHER READING

BEACH, Joseph Warren, *The Method of Henry James*, New Haven: Yale University Press, 1918; London: Oxford University Press, 1918. (Reprinted, Philadelphia: Albert Saifer, 1954.)

BEWLEY, Marius, *The Complex Fate*, London: Chatto & Windus, 1952; New York: Grove Press, 1954.

COVENEY, Peter, *The Image of Childhood. The Individual and Society: A Study of a Theme in English Literature.* Revised edition [of *Poor Monkey*, 1957], with an introduction by F. R. Leavis: Penguin, 1967.

FAHEY, P, '*What Maisie Knew*: Learning not to mind', *The Critical Review*, 14: Melbourne, 1971, pp. 96–108.

JAMES, Henry, *The Notebooks of Henry James*, ed. with introduction by F. O. Mattheissen and Kenneth B. Murdock, New York and London: Oxford University Press, 1947.

LEAVIS, F. R., '*What Maisie Knew*: A Disagreement by F. R. Leavis', in Marius Bewley, *The Complex Fate* (q.v.).

TANNER, Tony, *The Reign of Wonder: Naïvety and Reality in American Literature*: Cambridge, 1965.

WALSH, William, *The Use of Imagination*, London: Chatto & Windus, 1959; New York: Barnes & Noble, 1960.

WORDEN, Ward S., 'A Cut Version of *What Maisie Knew*', *American Literature*, xxiv (Jan. 1953), 493–504.

—— 'Henry James's *What Maisie Knew*: A Comparison with the Plans in *The Notebooks*', *P.M.L.A.* lxviii (June 1953), 371–83.

PREFACE

I RECOGNISE again, for the first of these three Tales, another instance of the growth of the "great oak" from the little acorn; since "What Maisie Knew" is at least a tree that spreads beyond any provision its small germ might on a first handling have appeared likely to make for it. The accidental mention had been made to me of the manner in which the situation of some luckless child of a divorced couple was affected, under my informant's eyes, by the remarriage of one of its parents—I forget which; so that, thanks to the limited desire for its company expressed by the step-parent, the law of its little life, its being entertained in rotation by its father and its mother, would n't easily prevail. Whereas each of these persons had at first vindictively desired to keep it from the other, so at present the re-married relative sought now rather to be rid of it—that is to leave it as much as possible, and beyond the appointed times and seasons, on the hands of the adversary; which malpractice, resented by the latter as bad faith, would of course be repaid and avenged by an equal treachery. The wretched infant was thus to find itself practically disowned, rebounding from racquet to racquet like a tennis-ball or a shuttlecock. This figure could but touch the fancy to the quick and strike one as the beginning of a story—a story commanding a great choice of developments. I recollect, however, promptly thinking that for a proper symmetry the second parent should marry too—which in the case named to me indeed would probably soon occur, and was in any case what the ideal of the situation required. The second step-parent would have but to be correspondingly incommoded by obligations to the offspring of a hated predecessor for the misfortune of the little victim to become altogether exemplary. The business would accordingly be sad enough, yet I am not sure its possibility of interest would so much have appealed to me had I not soon felt that the ugly facts, so stated or conceived, by no means constituted the whole appeal.

The light of an imagination touched by them could n't help therefore projecting a further ray, thanks to which it became rather quaintly clear that, not less than the chance of misery and of a degraded state, the chance of happiness and of an improved state might be here involved for the child, round about whom the complexity of life would thus turn to fineness, to richness—and indeed would have but so to turn for the small creature to be steeped in security and ease. Sketchily clustered even, these elements gave out that vague pictorial glow which forms the first appeal of a living "subject" to the painter's consciousness; but the glimmer became intense as I proceeded to a further analysis. The further analysis is for that matter almost always the torch of rapture and victory, as the artist's firm hand grasps and plays it—I mean, naturally, of the smothered rapture and the obscure victory, enjoyed and celebrated not in the street but before some innermost shrine; the odds being a hundred to one, in almost any connexion, that it does n't arrive by any easy first process at the *best* residuum of truth. That was the charm, sensibly, of the picture thus at first confusedly showing; the elements so could n't but flush, to their very surface, with some deeper depth of irony than the mere obvious. It lurked in the crude postulate like a buried scent; the more the attention hovered the more aware it became of the fragrance. To which I may add that the more I scratched the surface and penetrated, the more potent, to the intellectual nostril, became this virtue. At last, accordingly, the residuum, as I have called it, reached, I was in presence of the red dramatic spark that glowed at the core of my vision and that, as I gently blew upon it, burned higher and clearer. This precious particle was the *full* ironic truth—the most interesting item to be read into the child's situation. For satisfaction of the mind, in other words, the small expanding consciousness would have to be saved, have to become presentable as a register of impressions; and saved by the experience of certain advantages, by some enjoyed profit and some achieved confidence, rather than coarsened, blurred, sterilised, by ignorance and pain. This better state, in the young life, would reside in the exercise of a function other than that of disconcerting the selfishness of its parents —which was all that had on the face of the matter seemed reserved to it in the way of criticism applied to their rupture. The early relation would be exchanged for a later; instead of simply submitting to the inherited tie and the imposed complication, of suffering from them, our little wonder-working agent would create, without design, quite fresh elements of this order—contribute, that is, to the formation of a fresh tie, from which it would then (and for all the world as if through a small demonic foresight) proceed to derive great profit.

This is but to say that the light in which the vision so readily grew to a wholeness was that of a second marriage on both sides; the father having, in the freedom of divorce, but to take another wife, as well as the mother, under a like licence, another husband, for the case to begin, at least, to stand beautifully on its feet. There would be thus a perfect logic for what might come—come even with the mere attribution of a certain sensibility (if but a mere relative fineness) to either of the new parties. Say the prime cause making for the ultimate attempt to shirk on one side or the other, and better still if on both, a due share of the decreed burden should have been, after all, in each progenitor, a constitutional inaptitude for *any* burden, and a base intolerance of it: we should thus get a motive not requiring, but happily dispensing with, too particular a perversity in the stepparents. The child seen as creating by the fact of its forlornness a relation between its step-parents, the more intimate the better, dramatically speaking; the child, by the mere appeal of neglectedness and the mere consciousness of relief, weaving about, with the best faith in the world, the close web of sophistication; the child becoming a centre and pretext for a fresh system of misbehaviour, a system moreover of a nature to spread and ramify: *there* would be the "full" irony, there the promising theme into which the hint I had originally picked up would logically flower. No themes are so human as those that reflect for us, out of the confusion of life, the close connexion of bliss and bale, of the things that help with the things that hurt, so dangling before us for ever that bright hard medal, of so strange an alloy, one face of which is somebody's right and ease and the other somebody's pain and wrong. To live with all intensity and perplexity and felicity in its terribly mixed little world would thus be the part of my interesting small mortal; bringing people together who would be at least more correctly separate; keeping people separate who would be at least more correctly together; flourishing, to a degree, at the cost of many conventions and proprieties, even decencies; really keeping the torch of virtue alive in an air tending infinitely to smother it; really in short making confusion worse confounded by drawing some stray fragrance of an ideal across the scent of selfishness, by sowing on barren strands, through the mere fact of presence, the seed of the moral life.

All this would be to say, I at once recognised, that my light vessel of consciousness, swaying in such a draught, could n't be with verisimilitude a rude little boy; since, beyond the fact that little boys are never so "present," the sensibility of the female young is indubitably, for early youth, the greater, and my plan would call, on the part of my protagonist, for "no end" of sensibility. I might impute that amount of it without extravagance to a slip of a girl whose faculties should have been well shaken up; but I should have so to depend on its action to keep my story clear that I must be able to show it in all assurance as naturally intense. To this end I should have of course to suppose for my heroine dispositions originally promising, but above all I should have to invest her with perceptions easily and almost infinitely quickened. So handsomely fitted out, yet not in a manner too grossly to affront probability, she might well see me through the whole course of my design; which design, more and more attractive as I turned it over, and dignified by the most delightful difficulty, would be to make and to keep her so limited consciousness the very field of my picture while at the same time guarding with care the integrity of the objects represented. With the charm of this possibility, therefore, the project for "Maisie" rounded itself and loomed large— any subject looming large, for that matter, I am bound to add, from the moment one is ridden by the law of entire expression. I have already elsewhere noted, I think, that the memory of my own work preserves for me no theme that, at some moment or other of its development, and always only waiting for the right connexion or chance, has n't signally refused to remain humble, even (or perhaps all the more resentfully) when fondly selected for its conscious and hopeless humility. Once "out," like a house-dog of a temper above confinement, it defies the mere whistle, it roams, it hunts, it seeks out and "sees" life; it can be brought back but by hand and then only to take its futile thrashing. It was n't at any rate for an idea seen in the light I here glance at not to have due warrant of its value—how could the value of a scheme so finely workable *not* be great? The one presented register of the whole complexity would be the play of the child's confused and obscure notation of it, and yet the whole, as I say, should be unmistakeably, should be honourably there, seen through the faint intelligence, or at the least attested by the imponderable presence, and still advertising its sense.

I recall that my first view of this neat possibility was as the attaching problem of the picture restricted (while yet achieving, as I say, completeness and coherency) to what the child might be conceived to have *understood*—to have been able to interpret and appreciate. Further reflexion and experiment showed me my subject strangled in that extreme of rigour. The infant mind would at the best leave great gaps and voids; so that with a systematic surface possibly beyond reproach we should nevertheless fail of clearness of sense. I should have to stretch the matter to what my wondering witness materially and inevitably *saw*; a great deal of which quantity she either would n't understand at all or would quite misunderstand—and on those lines, only on those, my task would be prettily cut out. To that then I settled—to the question of giving it *all*, the whole situation surrounding her, but of giving it only through the occasions and connexions of her proximity and her attention; only as it might pass before her and appeal to her, as it might touch her and affect her, for better or worse, for perceptive gain or perceptive loss: so that we fellow witnesses, we not more invited but only more expert critics, should feel in strong possession of it. This would be, to begin with, a plan of absolutely definite and measurable application—that in itself always a mark of beauty; and I have been interested to find on re-perusal of the work that some such controlling grace successfully rules it. Nothing could be more "done," I think, in the light of its happiest intention; and this in spite of an appearance that at moments obscures my consistency. Small children have many more perceptions than they have terms to translate them; their vision is at any moment much richer, their apprehension even constantly stronger, than their prompt, their at all producible, vocabulary. Amusing therefore as it might at the first blush have seemed to restrict myself in this case to the terms as well as to the experience, it became at once plain that such an attempt would fail. Maisie's terms accordingly play their part—since her simpler conclusions quite depend on them; but our own commentary constantly attends and amplifies. This it is that on occasion, doubtless, seems to represent us as going so "behind" the facts of her spectacle as to exaggerate the activity of her relation to them. The difference here is but of a shade: it is her relation, her activity of spirit, that determines all our own concern—we simply take advantage of these things better than she herself. Only, even though it is her interest that mainly makes matters interesting for us, we inevitably note this in figures that are not yet at her command and that are nevertheless required whenever those aspects about her and those parts of her experience that she understands darken off into others that she rather tormentedly misses. All of which gave me a high firm logic to observe; supplied the force for which the straightener of almost any tangle is grateful while he labours, the sense of pulling at threads intrinsically worth it—strong enough and fine enough and entire enough.

Of course, beyond this, was another and well-nigh equal charm—equal in spite of its being almost independent of the acute constructional, the endless expressional question. This was the quite different question of the particular kind of truth of resistance I might be able to impute to my central figure—*some* intensity, some continuity of resistance being naturally of the essence of the subject. Successfully to resist (to resist, that is, the strain of observation and the assault of experience) what would that be, on the part of so young a person, but to remain fresh, and still fresh, and to have even a freshness to communicate?—the case being with Maisie to the end that she treats her friends to the rich little spectacle of objects embalmed in her wonder. She wonders, in other words, to the end, to the death—the death of her childhood, properly speaking; after which (with the inevitable shift, sooner or later, of her point of view) her situation will change and become another affair, subject to other measurements and with a new centre altogether. The particular reaction that will have led her to that point, and that it has been of an exquisite interest to study in her, will have spent itself; there will be another scale, another perspective, another horizon. Our business meanwhile therefore is to extract from her current reaction whatever it may be worth; and for that matter we recognise in it the highest exhibitional virtue. Truly, I reflect, if the theme had had no other beauty it would still have had this rare and distinguished one of its so expressing the variety of the child's values. She is not only the extra-ordinary "ironic centre" I have already noted; she has the wonderful importance of shedding a light far beyond any reach of her comprehension; of lending to poorer persons and things, by the mere fact of their being involved with her and by the special scale she creates for them, a precious element of dignity. I lose myself, truly, in appreciation of my theme on noting what she does by her "freshness" for appearances in themselves vulgar and empty enough. They become, as she deals with them, the stuff of poetry and tragedy and art; she has simply to wonder, as I say, about them, and they begin to have meanings, aspects, solidities, connexions—connexions with the "universal!"—that they could scarce have hoped for. Ida Farange alone, so to speak, or Beale alone, that is either of them otherwise connected— what intensity, what "objectivity" (the most developed degree of *being* anyhow thinkable for them) would they have? How would they repay at all the favour of our attention?

Maisie makes them portentous all by the play of her good faith, makes her mother above all, to my vision—unless I have wholly failed to render it—concrete, immense and awful; so that we get, for our profit, and get by an economy of process interesting in itself, the thoroughly pictured creature, the striking figured symbol. At two points in parti- cular, I seem to recognise, we enjoy at its maximum this effect of associational magic. The passage in which her father's terms of intercourse with the insinuating but so strange and unattractive lady whom he has had the detest- able levity to whisk her off to see late at night, is a signal example of the all but incalculable way in which interest may be constituted. The facts involved are that Beale Farange is ignoble, that the friend to whom he introduces his daughter is deplorable, and that from the commerce of the two, *as* the two merely, we would fain avert our heads. Yet the thing has but to become a part of the child's bewilderment for these small sterilities to drop from it and for the *scene* to emerge and prevail—vivid, special, wrought hard, to the hardness of the unforgettable; the scene that is exactly what Beale and Ida and Mrs. Cuddon, and even Sir Claude and Mrs. Beale, would never for a moment have succeeded in making their scant unredeemed importances—namely *appreciable*. I find another instance in the episode of Maisie's unprepared encounter, while walking in the Park with Sir Claude, of her mother and that beguiled attendant of her mother, the encouraging, the appealing "Captain," to whom this lady contrives to commit her for twenty minutes while she herself deals with the second husband. The human substance here would have seemed in advance well-nigh too poor for conversion, the three "mature" figures of too short a radiation, too stupid (*so* stupid it was for Sir Claude to have married Ida!) too vain, too thin, for any clear application; but promptly, immediately, the child's own importance, spreading and contagiously acting, has determined the *total* value otherwise. Nothing of course, meanwhile, is an older story to the observer of manners and the painter of life than the grotesque finality with which such terms as "painful," "unpleasant" and "disgusting" are often applied to his results; to that degree, in truth, that the free use of them as weightily conclusive again and again re-enforces his estimate of the critical sense of circles in which they artlessly flourish. Of course under that superstition I was punctually to have had read to me the lesson that the "mixing-up" of a child with anything unpleasant confessed itself an aggravation of the unpleasantness, and that nothing could well be more disgusting than to attribute to Maisie so intimate an "acquaintance" with the gross immoralities surrounding her.

The only thing to say of such lucidities is that, however one may have "discounted" in advance, and as once for all, their general radiance, one is disappointed if the hour for them, in the particular connexion, does n't strike—they so keep before us elements with which even the most sedate philosopher must always reckon. The painter of life has indeed work cut out for him when a considerable part of life offers itself in the guise of that sapience. The effort really to see and really to represent is no idle business in face of the *constant* force that makes for muddlement. The great thing is indeed that the muddled state too is one of the very sharpest of the realities, that it also has colour and form and character, has often in fact a broad and rich comicality, many of the signs and values of the appreciable. Thus it was to be, for example, I might gather, that the very principle of Maisie's appeal, her undestroyed freshness, in other words that vivacity of intelligence by which she indeed does vibrate in the infected air, indeed does flourish in her immoral world, may pass for a barren and senseless thing, or at best a negligible one. For nobody to whom life at large is *easily* interesting do the finer, the shyer, the more anxious small vibrations, fine and shy and anxious with the passion that precedes knowledge, succeed in being negligible: which is doubtless one of many reasons why the passage between the child and the kindly, friendly, ugly gentleman who, seated with her in Kensington Gardens under a spreading tree, positively answers to her for her mother as no one has ever answered, and so stirs her, filially and morally, as she has never been stirred, throws into highest relief, to my sense at least, the side on which the subject is strong, and becomes the type-passage—other advantages certainly aiding, as I may say—for the expression of its beauty. The active, contributive close-circling wonder, as I have called it, in which the child's identity is guarded and preserved, and which makes her case remarkable exactly by the weight of the tax on it, provides distinction for her, provides vitality and variety, through the operation of the tax—which would have done comparatively little for us had n't it been monstrous. A pity for us surely to have been deprived of this just reflexion. "Maisie" is of 1897.[1]

---

[1] The New York edition reads 1907.

WHAT MAISIE KNEW

THE litigation had seemed interminable and had in fact been complicated; but by the decision on the appeal the judgement of the divorce-court was confirmed as to the assignment of the child. The father, who, though bespattered from head to foot, had made good his case, was, in pursuance of this triumph, appointed to keep her: it was not so much that the mother's character had been more absolutely damaged as that the brilliancy of a lady's complexion (and this lady's, in court, was immensely remarked) might be more regarded as showing the spots. Attached, however, to the second pronouncement was a condition that detracted, for Beale Farange, from its sweetness—an order that he should refund to his late wife the twenty-six hundred pounds put down by her, as it was called, some three years before, in the interest of the child's maintenance and precisely on a proved understanding that he would take no proceedings: a sum of which he had had the administration and of which he could render not the least account. The obligation thus attributed to her adversary was no small balm to Ida's resentment; it drew a part of the sting from her defeat and compelled Mr. Farange perceptibly to lower his crest. He was unable to produce the money or to raise it in any way; so that after a squabble scarcely less public and scarcely more decent than the original shock of battle his only issue from his predicament was a compromise proposed by his legal advisers and finally accepted by hers.

His debt was by this arrangement remitted to him and the little girl disposed of in a manner worthy of the judgement-seat of Solomon. She was divided in two and the portions tossed impartially to the disputants. They would take her, in rotation, for six months at a time; she would spend half the year with each. This was odd justice in the eyes of those who still blinked in the fierce light projected from the tribunal—a light in which neither parent figured in the least as a happy example to youth and innocence. What was to have been expected on the evidence was the nomination, *in loco parentis*, of some proper third person, some respectable or at least some presentable friend. Apparently, however, the circle of the Faranges had been scanned in vain for any such ornament; so that the only solution finally meeting all the difficulties was, save that of sending Maisie to a Home, the partition of the tutelary office in the manner I have mentioned. There were more reasons for her parents to agree to it than there had ever been for them to agree to anything; and they now prepared with her help to enjoy the distinction that waits upon vulgarity sufficiently attested. Their rupture had resounded, and after being perfectly insignificant together they would be decidedly striking apart. Had they not produced an impression that warranted people in looking for appeals in the newspapers for the rescue of the little one—reverberation, amid à vociferous public, of the idea that some movement should be started or some benevolent person should come forward? A good lady came indeed a step or two: she was distantly related to Mrs. Farange, to whom she proposed that, having children and nurseries wound up and going, she should be allowed to take home the bone of contention and, by working it into her system, relieve at least one of the parents. This would make every time, for Maisie, after her inevitable six months with Beale, much more of a change.

"More of a change?" Ida cried. "Won't it be enough of a change for her to come from that low brute to the person in the world who detests him most?"

"No, because you detest him so much that you'll always talk to her about him. You'll keep him before her by perpetually abusing him."

Mrs. Farange stared. "Pray, then, am I to do nothing to counteract his villainous abuse of *me*?"

The good lady, for a moment, made no reply: her silence was a grim judgement of the whole point of view. "Poor little monkey!" she at last exclaimed; and the words were an epitaph for the tomb of Maisie's childhood. She was abandoned to her fate. What was clear to any spectator was that the only link binding her to either parent was this lamentable fact of her being a ready vessel for bitterness, a deep little porcelain cup in which biting acids could be mixed. They had wanted her not for any good they could do her, but for the harm they could, with her unconscious aid, do each other. She should serve their anger and seal their revenge, for husband and wife had been alike crippled by the heavy hand of justice, which in the last resort met on neither side their indignant claim to get, as they called it, everything. If each was only to get half this seemed to concede that neither was so base as the other pretended, or, to put it differently, offered them both as bad indeed, since they were only as good as each other. The mother had wished to prevent the father from, as she said, "so much as looking" at the child; the father's plea was that the mother's lightest touch was "simply contamination." These were the opposed principles in which Maisie was to be educated—she was to fit them together as she might. Nothing could have been more touching at first than her failure to suspect the ordeal that awaited her little unspotted soul. There were persons horrified to think what those in charge of it would combine to try to make of it: no one could conceive in advance that they would be able to make nothing ill.

This was a society in which for the most part people were occupied only with chatter, but the disunited couple had at last grounds for expecting a time of high activity. They girded their loins, they felt as if the quarrel had only begun. They felt indeed more married than ever, inasmuch as what marriage had mainly suggested to them was the unbroken opportunity to quarrel. There had been "sides" before, and there were sides as much as ever; for the sider too the prospect opened out, taking the pleasant form of a superabundance of matter for desultory conversation. The many friends of the Faranges drew together to differ about them; contradiction grew young again over teacups and cigars. Everybody was always assuring everybody of something very shocking, and nobody would have been jolly if nobody had been outrageous. The pair appeared to have a social attraction which failed merely as regards each other: it was indeed a great deal to be able to say for Ida that no one but Beale desired her blood, and for Beale that if he should ever have his eyes scratched out it would be only by his wife. It was generally felt, to begin with, that they were awfully good-looking—they had really not been analysed to a deeper residuum. They made up together for instance some twelve feet three of stature, and nothing was more discussed than the apportionment of this quantity. The sole flaw in Ida's beauty was a length and reach of arm conducive perhaps to her having so often beaten her ex-husband at billiards, a game in which she showed a superiority largely accountable, as she maintained, for the resentment finding expression in his physical violence. Billiards was her great accomplishment and the distinction her name always first produced the mention of. Notwithstanding some very long lines everything about her that might have been large and that in many women profited by the licence was, with a single exception, admired and cited for its smallness. The exception was her eyes, which might have been of mere regulation size, but which overstepped the modesty of nature; her mouth, on the other hand, was barely perceptible, and odds were freely taken as to the measurement of her waist. She was a person who, when she was out—and she was always out—produced everywhere a sense of having been seen often, the sense indeed of a kind of abuse of visibility, so that it would have been, in the usual places, rather vulgar to wonder at her. Strangers only did that; but they, to the amusement of the familiar, did it very much: it was an inevitable way of betraying an alien habit. Like her husband she carried clothes, carried them as a train carries passengers: people had been known to compare their taste and dispute about the accommodation they gave these articles, though inclining on the whole to the commendation of Ida as less overcrowded, especially with jewellery and flowers. Beale Farange had natural decorations, a kind of costume in his vast fair beard, burnished like a gold breastplate, and in the eternal glitter of the teeth that his long moustache had been trained not to hide and that gave him, in every possible situation, the look of the joy of life. He had been destined in his youth for diplomacy and momentarily attached, without a salary, to a legation which enabled him often to say "In *my* time in the East": but contemporary history had somehow had no use for him, had hurried past him and left him in perpetual Piccadilly. Every one knew what he had—only twenty-five hundred. Poor Ida, who had run through everything, had now nothing but her carriage and her paralysed uncle. This old brute, as he was called, was supposed to have a lot put away. The child was provided for, thanks to a crafty godmother, a defunct aunt of Beale's, who had left her something in such a manner that the parents could appropriate only the income.

I

THE child was provided for, but the new arrangement was inevitably confounding to a young intelligence intensely aware that something had happened which must matter a good deal and looking anxiously out for the effects of so great a cause. It was to be the fate of this patient little girl to see much more than she at first understood, but also even at first to understand much more than any little girl, however patient, had perhaps ever understood before. Only a drummer-boy in a ballad or a story could have been so in the thick of the fight. She was taken into the confidence of passions on which she fixed just the stare she might have had for images bounding across the wall in the slide of a magic-lantern. Her little world was phantasmagoric—strange shadows dancing on a sheet. It was as if the whole performance had been given for her—a mite of a half-scared infant in a great dim theatre. She was in short introduced to life with a liberality in which the selfishness of others found its account, and there was nothing to avert the sacrifice but the modesty of her youth.

Her first term was with her father, who spared her only in not letting her have the wild letters addressed to her by her mother: he confined himself to holding them up at her and shaking them, while he showed his teeth, and then amusing her by the way he chucked them, across the room, bang into the fire. Even at that moment, however, she had a scared anticipation of fatigue, a guilty sense of not rising to the occasion, feeling the charm of the violence with which the stiff unopened envelopes, whose big monograms—Ida bristled with monograms—she would have liked to see, were made to whizz, like dangerous missiles, through the air. The greatest effect of the great cause was her own greater importance, chiefly revealed to her in the larger freedom with which she was handled, pulled hither and thither and kissed, and the proportionately greater niceness she was obliged to show. Her features had somehow become prominent; they were so perpetually nipped by the gentlemen who came to see her father and the smoke of whose cigarettes went into her face. Some of these gentlemen made her strike matches and light their cigarettes; others, holding her on knees violently jolted, pinched the calves of her legs till she shrieked—her shriek was much admired—and reproached them with being toothpicks. The word stuck in her mind and contributed to her feeling from this time that she was deficient in something that would meet the general desire. She found out what it was: it was a congenital tendency to the production of a substance to which Moddle, her nurse, gave a short ugly name, a name painfully associated at dinner with the part of the joint that she did n't like. She had left behind her the time when she had no desires to meet, none at least save Moddle's, who, in Kensington Gardens, was always on the bench when she came back to see if she had been playing too far. Moddle's desire was merely that she should n't do that, and she met it so easily that the only spots in that long brightness were the moments of her wondering what would become of her if, on her rushing back, there should be no Moddle on the bench. They still went to the Gardens, but there was a difference even there; she was impelled perpetually to look at the legs of other children and ask her nurse if *they* were toothpicks. Moddle was terribly truthful; she always said: "Oh my dear, you'll not find such another pair as your own." It seemed to have to do with something else that Moddle often said: "You feel the strain—that's where it is; and you'll feel it still worse, you know."

Thus from the first Maisie not only felt it, but knew she felt it. A part of it was the consequence of her father's telling her he felt it too, and telling Moddle, in her presence, that she must make a point of driving that home. She was familiar, at the age of six, with the fact that everything had been changed on her account, everything ordered to enable him to give himself up to her. She was to remember always the words in which Moddle impressed upon her that he did so give himself: "Your papa wishes you never to forget, you know, that he has been dreadfully put about." If the skin on Moddle's face had to Maisie the air of being unduly, almost painfully, stretched, it never presented that appearance so much as when she uttered, as she often had occasion to utter, such words. The child wondered if they did n't make it hurt more than usual; but it was only after some time that she was able to attach to the picture of her father's sufferings, and more particularly to her nurse's manner about them, the meaning for which these things had waited. By the time she had grown sharper, as the gentlemen who had criticised her calves used to say, she found in her mind a collection of images and echoes to which meanings were attachable— images and echoes kept for her in the childish dusk, the dim closet, the high drawers, like games she was n't yet big enough to play. The great strain meanwhile was that of carrying by the right end the things her father said about her mother—things mostly indeed that Moddle, on a glimpse of them, as if they had been complicated toys or difficult books, took out of her hands and put away in the closet. A wonderful assortment of objects of this kind she was to discover there later, all tumbled up too with the things, shuffled into the same receptacle, that her mother had said about her father.

She had the knowledge that on a certain occasion which

every day brought nearer her mother would be at the door to take her away, and this would have darkened all the days if the ingenious Moddle had n't written on a paper in very big easy words ever so many pleasures that she would enjoy at the other house. These promises ranged from "a mother's fond love" to "a nice poached egg to your tea," and took by the way the prospect of sitting up ever so late to see the lady in question dressed, in silks and velvets and diamonds and pearls, to go out: so that it was a real support to Maisie, at the supreme hour, to feel how, by Moddle's direction, the paper was thrust away in her pocket and there clenched in her fist. The supreme hour was to furnish her with a vivid reminiscence, that of a strange outbreak in the drawing-room on the part of Moddle, who, in reply to something her father had just said, cried aloud: "You ought to be perfectly ashamed of yourself—you ought to blush, sir, for the way you go on!" The carriage, with her mother in it, was at the door; a gentleman who was there, who was always there, laughed out very loud; her father, who had her in his arms, said to Moddle: "My dear woman, I'll settle *you* presently!" —after which he repeated, showing his teeth more than ever at Maisie while he hugged her, the words for which her nurse had taken him up. Maisie was not at the moment so fully conscious of them as of the wonder of Moddle's sudden disrespect and crimson face; but she was able to produce them in the course of five minutes when, in the carriage, her mother, all kisses, ribbons, eyes, arms, strange sounds and sweet smells, said to her: "And did your beastly papa, my precious angel, send any message to your own loving mamma?" Then it was that she found the words spoken by her beastly papa to be, after all, in her little bewildered ears, from which, at her mother's appeal, they passed, in her clear shrill voice, straight to her little innocent lips. "He said I was to tell you, from him," she faithfully reported, "that you're a nasty horrid pig!"

II

IN that lively sense of the immediate which is the very air of a child's mind the past, on each occasion, became for her as indistinct as the future: she surrendered herself to the actual with a good faith that might have been touching to either parent. Crudely as they had calculated they were at first justified by the event: she was the little feathered shuttle-cock they could fiercely keep flying between them. The evil they had the gift of thinking or pretending to think of each other they poured into her little gravely-gazing soul as into a boundless receptacle, and each of them had doubtless the best conscience in the world as to the duty of teaching her the stern truth that should be her safeguard against the other She was at the age for which all stories are true and all conceptions are stories. The actual was the absolute, the present alone was vivid. The objurgation for instance launched in the carriage by her mother after she had at her father's bidding punctually performed was a missive that dropped into her memory with the dry rattle of a letter falling into a pillar-box. Like the letter it was, as part of the contents of a well-stuffed post-bag, delivered in due course at the right address. In the presence of these overflowings, after they had continued for a couple of years, the associates of either party sometimes felt that something should be done for what they called "the real good, don't you know?" of the child. The only thing done, however, in general, took place when it was sighingly remarked that she fortunately was n't all the year round where she happened to be at the awkward moment, and that, furthermore, either from extreme cunning or from extreme stupidity, she appeared not to take things in.

The theory of her stupidity, eventually embraced by her parents, corresponded with a great date in her small still life: the complete vision, private but final, of the strange office she filled. It was literally a moral revolution and accomplished in the depths of her nature. The stiff dolls on the dusky shelves began to move their arms and legs; old forms and phrases began to have a sense that frightened her. She had a new feeling, the feeling of danger; on which a new remedy rose to meet it, the idea of an inner self or, in other words, of concealment. She puzzled out with imperfect signs, but with a prodigious spirit, that she had been a centre of hatred and a messenger of insult, and that everything was bad because she had been employed to make it so. Her parted lips locked themselves with the determination to be employed no longer. She would forget everything, she would repeat nothing, and when, as a tribute to the successful application of her system, she began to be called a little idiot, she tasted a pleasure new and keen. When therefore, as she grew older, her parents in turn announced before her that she had grown shockingly dull, it was not from any real contraction of her little stream of life. She spoiled their fun, but she practically added to her own. She saw more and more; she saw too much. It was Miss Overmore, her first governess, who on a momentous occasion had sown the seeds of secrecy; sown them not by anything she said, but by a mere roll of those fine eyes which Maisie already admired. Moddle had become at this time, after alternations of residence of which the child had no clear record, an image faintly embalmed in the remembrance of hungry disappearances from the nursery and distressful lapses in the alphabet, sad embarrassments, in particular, when invited to recognise something her nurse described as "the important letter haitch." Miss Overmore, however hungry, never disappeared: this marked her somehow as of higher rank, and the character was confirmed by a prettiness that Maisie supposed to be extraordinary. Mrs. Farange had described her as almost too pretty, and some one had asked what that mattered so long as Beale was n't there. "Beale or no Beale," Maisie had heard her mother reply, "I take her because she 's a lady and yet awfully poor. Rather nice people, but there are seven sisters at home. What do people mean?"

Maisie did n't know what people meant, but she knew very soon all the names of all the sisters; she could say them off better than she could say the multiplication-table. She privately wondered moreover, though she never asked, about the awful poverty, of which her companion also never spoke. Food at any rate came up by mysterious laws; Miss Overmore never, like Moddle, had on an apron, and when she ate she held her fork with her little finger curled out. The child, who watched her at many moments, watched her particularly at that one. "I think you're lovely," she often said to her; even mamma, who was lovely too, had not such a pretty way with the fork. Maisie associated this showier presence with her now being "big," knowing of course that nursery-governesses were only for little girls who were not, as she said, "really" little. She vaguely knew, further, somehow, that the future was still bigger than she, and that a part of what made it so was the number of governesses lurking in it and ready to dart out. Everything that had happened when she was really little was dormant, everything but the positive certitude, bequeathed from afar by Moddle, that the natural way for a child to have her parents was separate and successive, like her mutton and her pudding or her bath and her nap.

"*Does* he know he lies?"—that was what she had vivaciously asked Miss Overmore on the occasion which was so suddenly to lead to a change in her life.

"Does he know—?" Miss Overmore stared; she had a stocking pulled over her hand and was pricking at it with a needle which she poised in the act. Her task was homely, but her movement, like all her movements, graceful.

"Why papa."

"That he 'lies'?"

"That's what mamma says I'm to tell him—'that he lies and he knows he lies.'" Miss Overmore turned very red, though she laughed out till her head fell back; then she pricked again at her muffled hand so hard that Maisie wondered how she could bear it. "*Am* I to tell him?" the child went on. It was then that her companion addressed her in the unmistakeable language of a pair of eyes of deep dark grey. "I can't say No," they replied as distinctly as possible;

"I can't say No, because I'm afraid of your mamma, don't you see? Yet how can I say Yes after your papa has been so kind to me, talking to me so long the other day, smiling and flashing his beautiful teeth at me the time we met him in the Park, the time when, rejoicing at the sight of us, he left the gentlemen he was with and turned and walked with us, stayed with us for half an hour?" Somehow in the light of Miss Overmore's lovely eyes that incident came back to Maisie with a charm it had n't had at the time, and this in spite of the fact that after it was over her governess had never but once alluded to it. On their way home, when papa had quitted them, she had expressed the hope that the child would n't mention it to mamma. Maisie liked her so, and had so the charmed sense of being liked by her, that she accepted this remark as settling the matter and wonderingly conformed to it. The wonder now lived again, lived in the recollection of what papa had said to Miss Overmore: "I've only to look at you to see you're a person I can appeal to for help to save my daughter." Maisie's ignorance of what she was to be saved from did n't diminish the pleasure of the thought that Miss Overmore was saving her. It seemed to make them cling together as in some wild game of "going round."

III

SHE was therefore all the more startled when her mother said to her in connexion with something to be done before her next migration: "You understand of course that she's not going with you."

Maisie turned quite faint. "Oh I thought she was."

"It does n't in the least matter, you know, what you think," Mrs. Farange loudly replied; "and you had better indeed for the future, miss, learn to keep your thoughts to yourself." This was exactly what Maisie had already learned, and the accomplishment was just the source of her mother's irritation. It was of a horrid little critical system, a tendency, in her silence, to judge her elders, that this lady suspected her, liking as she did, for her own part, a child to be simple and confiding. She liked also to hear the report of the whacks she administered to Mr. Farange's character, to his pretensions to peace of mind: the satisfaction of dealing them diminished when nothing came back. The day was at hand, and she saw it, when she should feel more delight in hurling Maisie at him than in snatching her away; so much so that her conscience winced under the acuteness of a candid friend who had remarked that the real end of all their tugging would be that each parent would try to make the little girl a burden to the other—a sort of game in which a fond mother clearly would n't show to advantage. The prospect of not showing to advantage, a distinction in which she held she had never failed, begot in Ida Farange an ill humour of which several persons felt the effect. She determined that Beale at any rate should feel it; she reflected afresh that in the study of how to be odious to him she must never give way. Nothing could incommode him more than not to get the good, for the child, of a nice female appendage who had clearly taken a fancy to her. One of the things Ida said to the appendage was that Beale's was a house in which no decent woman could consent to be seen. It was Miss Overmore herself who explained to Maisie that she had had a hope of being allowed to accompany her to her father's, and that this hope had been dashed by the way her mother took it. "She says that if I ever do such a thing as enter his service I must never expect to show my face in this house again. So I 've promised not to attempt to go with you. If I wait patiently till you come back here we shall certainly be together once more."

Waiting patiently, and above all waiting till she should come back there, seemed to Maisie a long way round—it reminded her of all the things she had been told, first and last, that she should have if she 'd be good and that in spite of her goodness she had never had at all. "Then who 'll take care of me at papa's?"

"Heaven only knows, my own precious!" Miss Overmore replied, tenderly embracing her. There was indeed no doubt that she was dear to this beautiful friend. What could have proved it better than the fact that before a week was out, in spite of their distressing separation and her mother's prohibition and Miss Overmore's scruples and Miss Overmore's promise, the beautiful friend had turned up at her father's? The little lady already engaged there to come by the hour, a fat dark little lady with a foreign name and dirty fingers, who wore, throughout, a bonnet that had at first given her a deceptive air, too soon dispelled, of not staying long, besides asking her pupil questions that had nothing to do with lessons, questions that Beale Farange himself, when two or three were repeated to him, admitted to be awfully low—this strange apparition faded before the bright creature who had braved everything for Maisie's sake. The bright creature told her little charge frankly what had happened— that she had really been unable to hold out. She had broken her vow to Mrs. Farange; she had struggled for three days and then had come straight to Maisie's papa and told him the simple truth. She adored his daughter; she could n't give her up; she'd make for her any sacrifice. On this basis it had been arranged that she should stay; her courage had been rewarded; she left Maisie in no doubt as to the amount of courage she had required. Some of the things she said made a particular impression on the child—her declaration for instance that when her pupil should get older she'd understand better just how "dreadfully bold" a young lady, to do exactly what she had done, had to be.

"Fortunately your papa appreciates it; he appreciates it *immensely*"—that was one of the things Miss Overmore also said, with a striking insistence on the adverb. Maisie herself was no less impressed with what this martyr had gone through, especially after hearing of the terrible letter that had come from Mrs. Farange. Mamma had been so angry that, in Miss Overmore's own words, she had loaded her with insult—proof enough indeed that they must never look forward to being together again under mamma's roof. Mamma's roof, however, had its turn, this time, for the child, of appearing but remotely contingent, so that, to reassure her, there was scarce a need of her companion's secret, solemnly confided—the probability there would be no going back to mamma at all. It was Miss Overmore's private conviction, and a part of the same communication, that if Mr. Farange's daughter would only show a really marked preference, she would be backed up by "public opinion" in holding on to him. Poor Maisie could scarcely grasp that incentive, but she could surrender herself to the day. She had conceived her first passion, and the object of it was her governess. It had n't been put to her, and she could n't, or at any rate she did n't, put it to herself, that she liked Miss Overmore better than she liked papa; but it would have sustained her under such an imputation to feel herself able to reply that papa too liked Miss Overmore exactly as much. He had particularly told her so. Besides she could easily see it.

IV

ALL this led her on, but it brought on her fate as well, the day when her mother would be at the door in the carriage in which Maisie now rode on no occasions but these. There was no question at present of Miss Overmore's going back with her: it was universally recognised that her quarrel with Mrs. Farange was much too acute. The child felt it from the first; there was no hugging nor exclaiming as that lady drove her away—there was only a frightening silence, unenlivened even by the invidious enquiries of former years, which culminated, according to its stern nature, in a still more frightening old woman, a figure awaiting her on the very doorstep. "You're to be under this lady's care," said her mother. "Take her, Mrs. Wix," she added, addressing the figure impatiently and giving the child a push from which Maisie gathered that she wished to set Mrs. Wix an example of energy. Mrs. Wix took her and, Maisie felt the next day, would never let her go. She had struck her at first, just after

Miss Overmore, as terrible; but something in her voice at the end of an hour touched the little girl in a spot that had never even yet been reached. Maisie knew later what it was, though doubtless she could n't have made a statement of it: these were things that a few days' talk with Mrs. Wix quite lighted up. The principal one was a matter-Mrs. Wix herself always immediately mentioned: she had had a little girl quite of her own, and the little girl had been killed'on the spot. She had had absolutely nothing else in all the world, and her affliction had broken her heart. It was comfortably established between them that Mrs. Wix's heart was broken. What Maisie felt was that she had been, with passion and anguish, a mother, and that this was something Miss Overmore was not, something (strangely, confusingly) that mamma was even less.

So it was that in the course of an extraordinarily short time she found herself as deeply absorbed in the image of the little dead Clara Matilda, who, on a crossing in the Harrow Road, had been knocked down and crushed by the cruellest of hansoms, as she had ever found herself in the family group made vivid by one of seven. "She's your little dead sister," Mrs. Wix ended by saying, and Maisie, all in a tremor of curiosity and compassion, addressed from that moment a particular piety to the small accepted acquisition. Somehow she was n't a real sister, but that only made her the more romantic. It contributed to this view of her that she was never to be spoken of in that character to any one else—least of all to Mrs. Farange, who would n't care for her nor recognise the relationship: it was to be just an unutterable and inexhaustible little secret with Mrs. Wix. Maisie knew everything about her that could be known, everything she had said or done in her little mutilated life, exactly how lovely she was, exactly how her hair was curled and her frocks were trimmed. Her hair came down far below her waist—it was of the most wonderful golden brightness, just as Mrs. Wix's own had been a long time before. Mrs. Wix's own was indeed very remarkable still, and Maisie had felt at first that she should never get on with it. It played a large part in the sad and strange appearance, the appearance as of a kind of greasy greyness, which Mrs. Wix had presented on the child's arrival. It had originally been yellow, but time had turned that elegance to ashes, to a turbid sallow unvenerable white. Still excessively abundant, it was dressed in a manner of which the poor lady appeared not yet to have recognised the supersession, with a glossy braid, like a large diadem, on the top of the head, and behind, at the nape of the neck, a dingy rosette like a large button. She wore glasses which, in humble reference to a divergent obliquity of vision, she called her straighteners, and a little ugly snuff-coloured dress trimmed with satin bands in the form of scallops and glazed with antiquity. The straighteners, she explained to Maisie, were put on for the sake of others, whom, as she believed, they helped to recognise the bearing, otherwise doubtful, of her regard; the rest of the melancholy garb could only have been put on for herself. With the added suggestion of her goggles it reminded her pupil of the polished shell or corslet of a horrid beetle. At first she had looked cross and almost cruel; but this impression passed away with the child's increased perception of her being in the eyes of the world a figure mainly to laugh at. She was as droll as a charade or an animal toward the end of the "natural history"—a person whom people, to make talk lively, described to each other and imitated. Every one knew the straighteners; every one knew the diadem and the button, the scallops and satin bands; every one, though Maisie had never betrayed her, knew even Clara Matilda.

It was on account of these things that mamma got her for such low pay, really for nothing: so much, one day when Mrs. Wix had accompanied her into the drawing-room and left her, the child heard one of the ladies she found there—a lady with eyebrows arched like skipping-ropes and thick black stitching, like ruled lines for musical notes on beautiful white gloves—announce to another. She knew governesses were poor; Miss Overmore was unmentionably and Mrs. Wix ever so publicly so. Neither this, however, nor the old brown frock nor the diadem nor the button, made a difference for Maisie in the charm put forth through everything, the charm of Mrs. Wix's conveying that somehow, in her ugliness and her poverty, she was peculiarly and soothingly safe; safer than any one in the world, than papa, than mamma, than the lady with the arched eyebrows; safer even, though so much less beautiful, than Miss Overmore, on whose loveliness, as she supposed it, the little girl was faintly conscious that one could n't rest with quite the same tucked-in and kissed-for-good-night feeling. Mrs. Wix was as safe as Clara Matilda, who was in heaven and yet, embarrassingly, also in Kensal Green, where they had been together to see her little huddled grave. It was from something in Mrs. Wix's tone, which in spite of caricature remained indescribable and inimitable, that Maisie, before her term with her mother was over, drew this sense of a support, like a breast-high banister in a place of "drops," that would never give way. If she knew her instructress was poor and queer she also knew she was not nearly so "qualified" as Miss Overmore, who could say lots of dates straight off (letting you hold the book yourself) state the position of Malabar, play six pieces without notes and, in a sketch, put in beautifully the trees and houses and difficult parts. Maisie herself could play more pieces than Mrs. Wix, who was moreover visibly ashamed of her houses and trees and could only, with the help of a smutty forefinger, of doubtful legitimacy in the field of art, do the smoke coming out of the chimneys.

They dealt, the governess and her pupil, in "subjects," but there were many the governess put off from week to week and that they never got to at all: she only used to say "We'll take that in its proper order." Her order was a circle as vast as the untravelled globe. She had not the spirit of adventure—the child could perfectly see how many subjects she was afraid of. She took refuge on the firm ground of fiction, through which indeed there curled the blue river of truth. She knew swarms of stories, mostly those of the novels she had read; relating them with a memory that never faltered and a wealth of detail that was Maisie's delight. They were all about love and beauty and countesses and wickedness. Her conversation was practically an endless narrative, a great garden of romance, with sudden vistas into her own life and gushing fountains of homeliness. These were the parts where they most lingered; she made the child take with her again every step of her long, lame course and think it beyond magic or monsters. Her pupil acquired a vivid vision of every one who had ever, in her phrase, knocked against her—some of them oh so hard!—every one literally but Mr. Wix, her husband, as to whom nothing was mentioned save that he had been dead for ages. He had been rather remarkably absent from his wife's career, and Maisie was never taken to see his grave.

V

THE second parting from Miss Overmore had been bad enough, but this first parting from Mrs. Wix was much worse. The child had lately been to the dentist's and had a term of comparison for the screwed-up intensity of the scene. It was dreadfully silent, as it had been when her tooth was taken out; Mrs. Wix had on that occasion grabbed her hand and they had clung to each other with the frenzy of their determination not to scream. Maisie, at the dentist's, had been heroically still, but just when she felt most anguish had become aware of an audible shriek on the part of her companion, a spasm of stifled sympathy. This was reproduced by the only sound that broke their supreme embrace when, a month later, the "arrangement," as her periodical uprootings were called, played the part of the horrible forceps. Embedded in Mrs. Wix's nature as her tooth had been socketed in her gum, the operation of extracting her would really have been a case for chloroform. It was a hug that fortunately left nothing to say, for the poor woman's want of words at such an hour seemed to fall in with her want of everything. Maisie's alternate parent, in the outermost vestibule—he liked the impertinence of crossing as much as that of his late wife's threshold—stood over them with his open watch and his still more open grin, while from the only corner of an eye on which something of Mrs. Wix's did n't impinge the child saw at the door a brougham in which Miss Overmore also waited. She remembered the difference when, six months before, she had been torn from the breast of that more spirited protectress. Miss Overmore, then also in the vestibule, but of course in the other one, had been thoroughly audible and voluble; her protest had rung out bravely and she had declared that something—her pupil did n't know exactly what—was a regular wicked shame. That had at the time dimly recalled to Maisie the far-away moment of Moddle's great outbreak: there seemed always to be "shames" connected in one way or another with her migrations. At present, while Mrs. Wix's arms tightened and the smell of her hair was strong, she further remembered how, in pacifying Miss Overmore, papa had made use of the words "you dear old duck!"—an expression which, by its oddity, had stuck fast in her young mind, having moreover a place well prepared for it there by what she knew of the governess whom she now always mentally characterised as the pretty one. She wondered whether this affection would be as great as before: that would at all events be the case with the prettiness Maisie could see in the face which showed brightly at the window of the brougham.

The brougham was a token of harmony, of the fine conditions papa would this time offer: he had usually come for her in a hansom, with a four-wheeler behind for the boxes. The four-wheeler with the boxes on it was actually there, but mamma was the only lady with whom she had ever been in a conveyance of the kind always of old spoken of by Moddle as a private carriage. Papa's carriage was, now that he had one, still more private, somehow, than mamma's; and when at last she found herself quite on top, as she felt, of its inmates and gloriously rolling away, she put to Miss Overmore, after another immense and talkative squeeze, a question of which the motive was a desire for information as to the continuity of a certain sentiment. "Did papa like you just the same while

I was gone?" she enquired—full of the sense of how markedly his favour had been established in her presence. She had bethought herself that this favour might, like her presence and as if depending on it, be only intermittent and for the season. Papa, on whose knee she sat, burst into one of those loud laughs of his that, however prepared she was, seemed always, like some trick in a frightening game, to leap forth and make her jump. Before Miss Overmore could speak he replied: "Why, you little donkey, when you're away what have I left to do but just to love her?" Miss Overmore hereupon immediately took her from him, and they had a merry little scrimmage over her of which Maisie caught the surprised perception in the white stare of an old lady who passed in a victoria. Then her beautiful friend remarked to her very gravely: "I shall make him understand that if he ever again says anything as horrid as that to you I shall carry you straight off and we'll go and live somewhere together and be good quiet little girls." The child could n't quite make out why her father's speech had been horrid, since it only expressed that appreciation which their companion herself had of old described as "immense." To enter more into the truth of the matter she appealed to him again directly, asked if in all those months Miss Overmore had n't been with him just as she had been before and just as she would be now. "Of course she has, old girl—where else could the poor dear be?" cried Beale Farange, to the still greater scandal of their companion, who protested that unless he straightway "took back" his nasty wicked fib it would be, this time, not only him she would leave, but his child too and his house and his tiresome troubles—all the impossible things he had succeeded in putting on her. Beale, under this frolic menace, took nothing back at all; he was indeed apparently on the point of repeating his extravagance, but Miss Overmore instructed her little charge that she was not to listen to his bad jokes: she was to understand that a lady could n't stay with a gentleman that way without some awfully proper reason.

Maisie looked from one of her companions to the other;

this was the freshest gayest start she had yet enjoyed, but she had a shy fear of not exactly believing them. "Well, what reason *is* proper?" she thoughtfully demanded.

"Oh a long-legged stick of a tomboy: there's none so good as that." Her father enjoyed both her drollery and his own and tried again to get possession of her—an effort deprecated by their comrade and leading again to something of a public scuffle. Miss Overmore declared to the child that she had been all the while with good friends; on which Beale Farange went on: "She means good friends of mine, you know—tremendous friends of mine. There has been no end of *them* about—that I *will* say for her!" Maisie felt bewildered and was afterwards for some time conscious of a vagueness, just slightly embarrassing, as to the subject of so much amusement and as to where her governess had really been. She did n't feel at all as if she had been seriously told, and no such feeling was supplied by anything that occurred later. Her embarrassment, of a precocious instinctive order, attached itself to the idea that this was another of the matters it was not for her, as her mother used to say, to go into. Therefore, under her father's roof during the time that followed, she made no attempt to clear up her ambiguity by an ingratiating way with housemaids; and it was an odd truth that the ambiguity itself took nothing from the fresh pleasure promised her by renewed contact with Miss Overmore. The confidence looked for by that young lady was of the fine sort that explanation can't improve, and she herself at any rate was a person superior to any confusion. For Maisie moreover concealment had never necessarily seemed deception; she had grown up among things as to which her foremost knowledge was that she was never to ask about them. It was far from new to her that the questions of the small are the peculiar diversion of the great: except the affairs of her doll Lisette there had scarcely ever been anything at her mother's that was explicable with a grave face. Nothing was so easy to her as to send the ladies who gathered there off into shrieks, and she might have practised upon them largely if she had been of a more calculating turn. Everything had something behind it: life was like a long, long corridor with rows of closed doors. She had learned that at these doors it was wise not to knock—this seemed to produce from within such sounds of derision. Little by little, however, she understood more, for it befell that she was enlightened by Lisette's questions, which reproduced the effect of her own upon those for whom she sat in the very darkness of Lisette. Was she not herself convulsed by such innocence? In the presence of it she often imitated the shrieking ladies. There were at any rate things she really could n't tell even a French doll. She could only pass on her lessons and study to produce on Lisette the impression of having mysteries in her life, wondering the while whether she succeeded in the air of shading off, like her mother, into the unknowable. When the reign of Miss Overmore followed that of Mrs. Wix she took a fresh cue, emulating her governess and bridging over the interval with the simple expectation of trust. Yes, there were matters one could n't "go into" with a pupil. There were for instance days when, after prolonged absence, Lisette, watching her take off her things, tried hard to discover where she had been. Well, she discovered a little, but never discovered all. There was an occasion when, on her, being particularly indiscreet, Maisie replied to her—and precisely about the motive of a disappearance—as she, Maisie, had once been replied to by Mrs. Farange: "Find out for yourself!" She mimicked her mother's sharpness, but she was rather ashamed afterwards, though as to whether of the sharpness or of the mimicry was not quite clear.

VI

SHE became aware in time that this phase would n't have shone by lessons, the care of her education being now only one of the many duties devolving on Miss Overmore; a devolution as to which she was present at various passages between that lady and her father—passages significant, on either side, of dissent and even of displeasure. It was gathered by the child on these occasions that there was something in the situation for which her mother might "come down" on them all, though indeed the remark, always dropped by her father, was greeted on his companion's part with direct contradiction. Such scenes were usually brought to a climax by Miss Overmore's demanding, with more asperity than she applied to any other subject, in what position under the sun such a person as Mrs. Farange would find herself for coming down. As the months went on the little girl's interpretations thickened, and the more effectually that this stretch was the longest she had known without a break. She got used to the idea that her mother, for some reason, was in no hurry to reinstate her: that idea was forcibly expressed by her father whenever Miss Overmore, differing and decided, took him up on the question, which he was always putting forward, of the urgency of sending her to school. For a governess Miss Overmore differed surprisingly; far more for instance than would have entered into the bowed head of Mrs. Wix. She observed to Maisie many times that she was quite conscious of not doing her justice, and that Mr. Farange equally measured and equally lamented this deficiency. The reason of it was that she had mysterious responsibilities that interfered—responsibilities, Miss Overmore intimated, to Mr. Farange himself and to the friendly noisy little house and those who came there. Mr. Farange's remedy for every inconvenience was that the child should be put at school—there were such lots of splendid schools, as everybody knew, at Brighton and all over the place. That, however, Maisie learned, was just what would bring her mother down: from the moment he should delegate to others the housing of his little charge he had n't a leg to stand on before the law. Did n't he keep her away from her mother precisely because Mrs. Farange was one of these others?

There was also the solution of a second governess, a young person to come in by the day and really do the work; but to this Miss Overmore would n't for a moment listen, arguing against it with great public relish and wanting to know from all comers—she put it even to Maisie herself—if they did n't see how frightfully it would give her away. "What am I supposed to be at all, don't you see, if I'm not here to look after her?" She was in a false position and so freely and loudly called attention to it that it seemed to become almost a source of glory. The way out of it of course was just to do her plain duty; but that was unfortunately what, with his excessive, his exorbitant demands on her, which every one indeed appeared quite to understand, he practically, he selfishly prevented. Beale Farange, for Miss Overmore, was now never anything but "he," and the house was as full as ever of lively gentlemen with whom, under that designation, she chaffingly talked about him. Maisie meanwhile, as a subject of familiar gossip on what was to be done with her, was left so much to herself that she had hours of wistful thought of the large loose discipline of Mrs. Wix; yet she none the less held it under her father's roof a point of superiority that none of his visitors were ladies. It added to this odd security that she had once heard a gentleman say to him as if it were a great joke and in obvious reference to Miss Overmore: "Hanged if she'll let another woman come near you—hanged if she ever will. She'd let fly a stick at her as they do at a strange cat!" Maisie greatly preferred gentlemen as inmates in spite of their also having their way—louder but sooner over—of laughing out at her. They pulled and pinched, they teased and tickled her; some of them even, as they termed it, shied things at her, and all of them thought it funny to call her by names having no resemblance to her own. The ladies on the other hand addressed her as "You poor pet" and scarcely touched her even to kiss her. But it was of the ladies she was most afraid.

She was now old enough to understand how disproportionate a stay she had already made with her father; and also old enough to enter a little into the ambiguity attending this excess, which oppressed her particularly whenever the question had been touched upon in talk with her governess. "Oh you need n't worry: she does n't care!" Miss Overmore had often said to her in reference to any fear that her mother might resent her prolonged detention. "She has other people than poor little *you* to think about, and has gone abroad with them; so you need n't be in the least afraid she 'll stickle this time for her rights." Maisie knew Mrs. Farange had gone abroad, for she had had weeks and weeks before a letter from her beginning "My precious pet" and taking leave of her for an indeterminate time; but she had not seen in it a renunciation of hatred or of the writer's policy of asserting herself, for the sharpest of all her impressions had been that there was nothing her mother would ever care so much about as to torment Mr. Farange. What at last, however, was in this connexion bewildering and a little frightening was the dawn of a suspicion that a better way had been found to torment Mr. Farange than to deprive him of his periodical burden. This was the question that worried our young lady and that Miss Overmore's confidences and the frequent observations of her employer only rendered more mystifying. It was a contradiction that if Ida had now a fancy for waiving the rights she had originally been so hot about her late husband should n't jump at the monopoly for which he had also in the first instance so fiercely fought; but when Maisie, with a subtlety beyond her years, sounded this new ground her main success was in hearing her mother more freshly abused. Miss Overmore had up to now rarely deviated from a decent reserve, but the day came when she expressed herself with a vividness not inferior to Beale's own on the subject of the lady who had fled to the Continent to wriggle out of her job. It would serve this lady right, Maisie gathered, if that contract, in the shape of an overgrown and underdressed daughter, should be shipped straight out to her and landed at her feet in the midst of scandalous excesses.

The picture of these pursuits was what Miss Overmore took refuge in when the child tried timidly to ascertain if her father were disposed to feel he had too much of her. She evaded the point and only kicked up all round it the dust of Ida's heartlessness and folly, of which the supreme proof, it appeared, was the fact that she was accompanied on her journey by a gentleman whom, to be painfully plain on it, she had—well, "picked up." The only terms on which, unless they were married, ladies and gentlemen might, as Miss Overmore expressed it, knock about together, were the terms on which she and Mr. Farange had exposed themselves to possible misconception. She had indeed, as has been noted, often explained this before, often said to Maisie: "I don't know what in the world, darling, your father and I should do without you, for you just make the difference, as I've told you, of keeping us perfectly proper." The child took in the office it was so endearingly presented to her that she performed a comfort that helped her to a sense of security even in the event of her mother's giving her up. Familiar as she had grown with the fact of the great alternative to the proper, she felt in her governess and her father a strong reason for not emulating that detachment. At the same time she had heard somehow of little girls—of exalted rank, it was true—whose education was carried on by instructors of the other sex, and she knew that if she were at school at Brighton it would be thought an advantage to her to be more or less in the hands of masters. She turned these things over and remarked to Miss Overmore that if she should go to her mother perhaps the gentleman might become her tutor.

"The gentleman?" The proposition was complicated enough to make Miss Overmore stare.

"The one who's with mamma. Might n't that make it right—as right as your being my governess makes it for you to be with papa?"

Miss Overmore considered; she coloured a little; then she embraced her ingenious friend. "You're too sweet! I'm a *real* governess."

"And could n't he be a real tutor?"

"Of course not. He's ignorant and bad."

"Bad—?" Maisie echoed with wonder.

Her companion gave a queer little laugh at her tone. "He's ever so much younger—" But that was all.

"Younger than you?"

Miss Overmore laughed again; it was the first time Maisie had seen her approach so nearly to a giggle. "Younger than—

no matter whom. I don't know anything about him and don't want to," she rather inconsequently added. "He's not my sort, and I'm sure, my own darling, he's not yours." And she repeated the free caress into which her colloquies with Maisie almost always broke and which made the child feel that *her* affection at least was a gage of safety. Parents had come to seem vague, but governesses were evidently to be trusted. Maisie's faith in Mrs. Wix for instance had suffered no lapse from the fact that all communication with her had temporarily dropped. During the first weeks of their separation Clara Matilda's mamma had repeatedly and dolefully written to her, and Maisie had answered with an enthusiasm controlled only by orthographical doubts; but the correspondence had been duly submitted to Miss Overmore, with the final effect of its not suiting her. It was this lady's view that Mr. Farange would n't care for it at all, and she ended by confessing—since her pupil pushed her—that she did n't care for it herself. She was furiously jealous, she said; and that weakness was but a new proof of her disinterested affection. She pronounced Mrs. Wix's effusions moreover illiterate and unprofitable; she made no scruple of declaring it monstrous that a woman in her senses should have placed the formation of her daughter's mind in such ridiculous hands. Maisie was well aware that the proprietress of the old brown dress and the old odd headgear was lower in the scale of "form" than Miss Overmore; but it was now brought home to her with pain that she was educationally quite out of the question. She was buried for the time beneath a conclusive remark of her critic's: "She's really beyond a joke!" This remark was made as that charming woman held in her hand the last letter that Maisie was to receive from Mrs. Wix; it was fortified by a decree proscribing the preposterous tie. "Must I then write and tell her?" the child bewilderedly asked: she grew pale at the dreadful things it appeared involved for her to say. "Don't dream of it, my dear—*I'll* write: you may trust me!" cried Miss Overmore; who indeed wrote to such purpose that a hush in which you could have heard a pin drop descended

upon poor Mrs. Wix. She gave for weeks and weeks no sign
whatever of life: it was as if she had been as effectually disposed of by Miss Overmore's communication as her little
girl, in the Harrow Road, had been disposed of by the
terrible hansom. Her very silence became after this one of the
largest elements of Maisie's consciousness; it proved a warm
and habitable air, into which the child penetrated further
than she dared ever to mention to her companions. Somewhere in the depths of it the dim straighteners were fixed
upon her; somewhere out of the troubled little current Mrs.
Wix intensely waited.

VII

IT quite fell in with this intensity that one day, on returning
from a walk with the housemaid, Maisie should have found
her in the hall, seated on the stool usually occupied by the
telegraph-boys who haunted Beale Farange's door and
kicked their heels while, in his room, answers to their mis-
sives took form with the aid of smoke-puffs and growls. It
had seemed to her on their parting that Mrs. Wix had
reached the last limits of the squeeze, but she now felt those
limits to be transcended and that the duration of her visitor's
hug was a direct reply to Miss Overmore's veto. She understood in a flash how the visit had come to be possible—that
Mrs. Wix, watching her chance, must have slipped in under
protection of the fact that papa, always tormented in spite of
arguments with the idea of a school, had, for a three days'
excursion to Brighton, absolutely insisted on the attendance
of her adversary. It was true that when Maisie explained their
absence and their important motive Mrs. Wix wore an
expression so peculiar that it would only have had its origin
in surprise. This contradiction indeed peeped out only to
vanish, for at the very moment that, in the spirit of it, she
threw herself afresh upon her young friend a hansom crested
with neat luggage rattled up to the door and Miss Overmore

bounded out. The shock of her encounter with Mrs. Wix was less violent than Maisie had feared on seeing her and did n't at all interfere with the sociable tone in which, under her rival's eyes, she explained to her little charge that she had returned, for a particular reason, a day sooner than she first intended. She had left papa—in such nice lodgings—at Brighton; but he would come back to his dear little home on the morrow. As for Mrs. Wix, papa's companion supplied Maisie in later converse with the right word for the attitude of this personage: Mrs. Wix "stood up" to her in a manner that the child herself felt at the time to be astonishing. This occurred indeed after Miss Overmore had so far raised her interdict as to make a move to the dining-room, where, in the absence of any suggestion of sitting down, it was scarcely more than natural that even poor Mrs. Wix should stand up. Maisie at once enquired if at Brighton, this time, anything had come of the possibility of a school; to which, much to her surprise, Miss Overmore, who had always grandly repudiated it, replied after an instant, but quite as if Mrs. Wix were not there:

"It may be, darling, that something *will* come. The objection, I must tell you, has been quite removed."

At this it was still more startling to hear Mrs. Wix speak out with great firmness. "I don't think, if you'll allow me to say so, that there's any arrangement by which the objection *can* be 'removed.' What has brought me here to-day is that I've a message for Maisie from dear Mrs. Farange."

The child's heart gave a great thump. "Oh mamma's come back?"

"Not yet, sweet love, but she's coming," said Mrs. Wix, "and she has—most thoughtfully, you know—sent me on to prepare you."

"To prepare her for what, pray?" asked Miss Overmore, whose first smoothness began, with this news, to be ruffled.

Mrs. Wix quietly applied her straighteners to Miss Overmore's flushed beauty. "Well, miss, for a very important communication."

"Can't dear Mrs. Farange, as you so oddly call her, make

her communications directly? Can't she take the trouble to write to her only daughter?" the younger lady demanded. "Maisie herself will tell you that it's months and months since she has had so much as a word from her."

"Oh but I've written to mamma!" cried the child as if this would do quite as well.

"That makes her treatment of you all the greater scandal," the governess in possession promptly declared.

"Mrs. Farange is too well aware," said Mrs. Wix with sustained spirit, "of what becomes of her letters in this house."

Maisie's sense of fairness hereupon interposed for her visitor. "You know, Miss Overmore, that papa does n't like everything of mamma's."

"No one likes, my dear, to be made the subject of such language as your mother's letters contain. They were not fit for the innocent child to see," Miss Overmore observed to Mrs. Wix.

"Then I don't know what you complain of, and she's better without them. It serves every purpose that I'm in Mrs. Farange's confidence."

Miss Overmore gave a scornful laugh. "Then you must be mixed up with some extraordinary proceedings!"

"None so extraordinary," cried Mrs. Wix, turning very pale, "as to say horrible things about the mother to the face of the helpless daughter!"

"Things not a bit more horrible, I think," Miss Overmore returned, "than those you, madam, appear to have come here to say about the father!"

Mrs. Wix looked for a moment hard at Maisie, and then, turning again to this witness, spoke with a trembling voice. "I came to say nothing about him, and you must excuse Mrs. Farange and me if we're not so above all reproach as the companion of his travels."

The young woman thus described stared at the apparent breadth of the description—she needed a moment to take it in. Maisie, however, gazing solemnly from one of the disputants to the other, noted that her answer, when it came, perched upon smiling lips. "It will do quite as well, no

41

doubt, if you come up to the requirements of the companion of Mrs. Farange's!"

Mrs. Wix broke into a queer laugh; it sounded to Maisie an unsuccessful imitation of a neigh. "That's just what I'm here to make known—how perfectly the poor lady comes up to them herself." She held up her head at the child. "You must take your mamma's message, Maisie, and you must feel that her wishing me to come to you with it this way is a great proof of interest and affection. She sends you her particular love and announces to you that she's engaged to be married to Sir Claude."

"Sir Claude?" Maisie wonderingly echoed. But while Mrs. Wix explained that this gentleman was a dear friend of Mrs. Farange's, who had been of great assistance to her in getting to Florence and in making herself comfortable there for the winter, she was not too violently shaken to perceive her old friend's enjoyment of the effect of this news on Miss Overmore. That young lady opened her eyes very wide; she immediately remarked that Mrs. Farange's marriage would of course put an end to any further pretension to take her daughter back. Mrs. Wix enquired with astonishment why it should do anything of the sort, and Miss Overmore gave as an instant reason that it was clearly but another dodge in a system of dodges. She wanted to get out of the bargain: why else had she now left Maisie on her father's hands weeks and weeks beyond the time about which she had originally made such a fuss? It was vain for Mrs. Wix to represent—as she speciously proceeded to do—that all this time would be made up as soon as Mrs. Farange returned: she, Miss Overmore, knew nothing, thank heaven, about her confederate, but was very sure any person capable of forming that sort of relation with the lady in Florence would easily agree to object to the presence in his house of the fruit of a union that his dignity must ignore. It was a game like another, and Mrs. Wix's visit was clearly the first move in it. Maisie found in this exchange of asperities a fresh incitement to the unformulated fatalism in which her sense of her own career had long since taken refuge; and it was the beginning for her of a deeper prevision that, in spite of Miss Overmore's brilliancy and Mrs. Wix's passion, she should live to see a change in the nature of the struggle she appeared to have come into the world to produce. It would still be essentially a struggle, but its object would now be *not* to receive her.

Mrs. Wix, after Miss Overmore's last demonstration, addressed herself wholly to the little girl, and, drawing from the pocket of her dingy old pelisse a small flat parcel, removed its envelope and wished to know if *that* looked like a gentleman who would n't be nice to everybody—let alone to a person he would be so sure to find so nice. Mrs. Farange, in the candour of new-found happiness, had enclosed a "cabinet" photograph of Sir Claude, and Maisie lost herself in admiration of the fair smooth face, the regular features, the kind eyes, the amiable air, the general glossiness and smartness of her prospective stepfather—only vaguely puzzled to suppose herself now with two fathers at once. Her researches had hitherto indicated that to incur a second parent of the same sex you had usually to lose the first. "*Is n't* he sympathetic?" asked Mrs. Wix, who had clearly, on the strength of his charming portrait, made up her mind that Sir Claude promised her a future. "You can see, I hope," she added with much expression, "that *he's* a perfect gentleman!" Maisie had never before heard the word "sympathetic" applied to anybody's face; she heard it with pleasure and from that moment it agreeably remained with her. She testified moreover to the force of her own perception in a small soft sigh of response to the pleasant eyes that seemed to seek her acquaintance, to speak to her directly. "He's quite lovely!" she declared to Mrs. Wix. Then eagerly, irrepressibly, as she still held the photograph and Sir Claude continued to fraternise, "Oh can't I keep it?" she broke out. No sooner had she done so than she looked up from it at Miss Overmore: this was with the sudden instinct of appealing to the authority that had long ago impressed on her that she must n't ask for things. Miss Overmore, to her surprise, looked distant and rather odd, hesitating and giving her time to turn again to Mrs. Wix. Then Maisie saw that lady's long face lengthen; it was stricken and almost scared, as if her young friend really expected more of her than she had to give. The photograph was a possession that, direly denuded, she clung to, and there was a momentary struggle between her fond clutch of it and her capability of every sacrifice for her precarious pupil. With the acuteness of her years, however, Maisie saw that her own avidity would triumph, and she held out the picture to Miss Overmore as if she were quite proud of her mother. "Isn't he just lovely?" she demanded while poor Mrs. Wix hungrily wavered, her straighteners largely covering it and her pelisse gathered about her with an intensity that strained its ancient seams.

"It was to *me*, darling," the visitor said, "that your mamma so generously sent it; but of course if it would give you particular pleasure—" She faltered, only gasping her surrender.

Miss Overmore continued extremely remote. "If the photograph's your property, my dear, I shall be happy to oblige you by looking at it on some future occasion. But you must excuse me if I decline to touch an object belonging to Mrs. Wix."

That lady had by this time grown very red. "You might as well see him this way, miss," she retorted, "as you certainly never will, I believe, in any other! Keep the pretty picture, by all means, my precious," she went on: "Sir Claude will be happy himself, I dare say, to give me one with a kind inscription." The pathetic quaver of this brave boast was not lost on Maisie, who threw herself so gratefully on the speaker's neck that, when they had concluded their embrace, the public tenderness of which, she felt, made up for the sacrifice she imposed, their companion had had time to lay a quick hand on Sir Claude and, with a glance at him or not, whisk him effectually out of sight. Released from the child's arms Mrs. Wix looked about for the picture; then she fixed Miss Overmore with a hard dumb stare; and finally, with her eyes on the little girl again, achieved the grimmest of smiles. "Well, nothing matters, Maisie, because there's another thing your mamma wrote about. She has made sure of me." Even after her loyal hug Maisie felt a bit of a sneak as she glanced at Miss Overmore for permission to understand this. But Mrs. Wix left them in no doubt of what it meant. "She has definitely engaged me—for her return and for yours. Then you'll see for yourself." Maisie, on the spot, quite believed she should; but the prospect was suddenly thrown into confusion by an extraordinary demonstration from Miss Overmore.

"Mrs. Wix," said that young lady, "has some undiscoverable reason for regarding your mother's hold on you as strengthened by the fact that she's about to marry. I wonder then—on that system—what our visitor will say to your father's."

Miss Overmore's words were directed to her pupil, but her face, lighted with an irony that made it prettier even than ever before, was presented to the dingy figure that had stiffened itself for departure. The child's discipline had been bewildering—it had ranged freely between the prescription that she was to answer when spoken to and the experience of lively penalties on obeying that prescription. This time, nevertheless, she felt emboldened for risks; above all as something portentous seemed to have leaped into her sense of the relations of things. She looked at Miss Overmore much as she had a way of looking at persons who treated her to "grown up" jokes. "Do you mean papa's hold on me—do you mean *he's* about to marry?"

"Papa's not about to marry—papa *is* married, my dear. Papa was married the day before yesterday at Brighton." Miss Overmore glittered more gaily; meanwhile it came over Maisie, and quite dazzlingly, that her "smart" governess was a bride. "He's my husband, if you please, and I'm his little wife. So *now* we'll see who's your little mother!" She caught her pupil to her bosom in a manner that was not to be outdone by the emissary of her predecessor, and a few moments later, when things had lurched back into their places, that poor lady, quite defeated of the last word, had soundlessly taken flight.

VIII

AFTER Mrs. Wix's retreat Miss Overmore appeared to recognise that she was not exactly in a position to denounce Ida Farange's second union; but she drew from a table-drawer the photograph of Sir Claude, and, standing there before Maisie, studied it at some length.

"Is n't he beautiful?" the child ingenuously asked.

Her companion hesitated. "No—he's horrid," she, to Maisie's surprise, sharply returned. But she debated another minute, after which she handed back the picture. It appeared to Maisie herself to exhibit a fresh attraction, and she was troubled, having never before had occasion to differ from her lovely friend. So she only could ask what, such being the case, she should do with it: should she put it quite away—where it would n't be there to offend? On this Miss Overmore again cast about; after which she said unexpectedly: "Put it on the schoolroom mantelpiece."

Maisie felt a fear. "Won't papa dislike to see it there?"

"Very much indeed; but that won't matter *now*." Miss Overmore spoke with peculiar significance and to her pupil's mystification.

"On account of the marriage?" Maisie risked.

Miss Overmore laughed, and Maisie could see that in spite of the irritation produced by Mrs. Wix she was in high spirits. "Which marriage do you mean?"

With the question put to her it suddenly struck the child she did n't know, so that she felt she looked foolish. So she took refuge in saying: "Shall *you* be different—?" This was a full implication that the bride of Sir Claude would be.

"As your father's wedded wife? Utterly!" Miss Overmore replied. And the difference began of course in her being addressed, even by Maisie, from that day and by her particular request, as Mrs. Beale. It was there indeed principally that it ended, for except that the child could reflect that she should presently have four parents in all, and also that at

46

the end of three months the staircase, for a little girl hanging over banisters, sent up the deepening rustle of more elaborate advances, everything made the same impression as before. Mrs. Beale had very pretty frocks, but Miss Overmore's had been quite as good, and if papa was much fonder of his second wife than he had been of his first Maisie had foreseen that fondness, had followed its development almost as closely as the person more directly involved. There was little indeed in the commerce of her companions that her precocious experience could n't explain, for if they struck her as after all rather deficient in that air of the honeymoon of which she had so often heard—in much detail, for instance, from Mrs. Wix—it was natural to judge the circumstance in the light of papa's proved disposition to contest the empire of the matrimonial tie. His honeymoon, when he came back from Brighton—not on the morrow of Mrs. Wix's visit, and not, oddly, till several days later—his honeymoon was perhaps perceptibly tinged with the dawn of a later stage of wedlock. There were things dislike of which, as the child knew it, would n't matter to Mrs. Beale now, and their number increased so that such a trifle as his hostility to the photograph of Sir Claude quite dropped out of view. This pleasing object found a conspicuous place in the schoolroom, which in truth Mr. Farange seldom entered and in which silent admiration formed, during the time I speak of, almost the sole scholastic exercise of Mrs. Beale's pupil.

Maisie was not long in seeing just what her stepmother had meant by the difference she should show in her new character. If she was her father's wife she was not her own governess, and if her presence had had formerly to be made regular by the theory of a humble function she was now on a footing that dispensed with all theories and was inconsistent with all servitude. That was what she had meant by the drop of the objection to a school; her small companion was no longer required at home as—it was Mrs. Beale's own amusing word—a little duenna. The argument against a successor to Miss Overmore remained: it was composed frankly of the fact, of which Mrs. Beale granted the full absurdity, that she was too awfully fond of her stepdaughter to bring herself to see her in vulgar and mercenary hands. The note of this particular danger emboldened Maisie to put in a word for Mrs. Wix, the modest measure of whose avidity she had taken from the first; but Mrs. Beale disposed afresh and effectually of a candidate who would be sure to act in some horrible and insidious way for Ida's interest and who moreover was personally loathsome and as ignorant as a fish. She made also no more of a secret of the awkward fact that a good school would be hideously expensive, and of the further circumstance, which seemed to put an end to everything, that when it came to the point papa, in spite of his previous clamour, was really most nasty about paying. "Would you believe," Mrs. Beale confidentially asked of her little charge, "that he says I'm a worse expense than ever, and that a daughter and a wife together are really more than he can afford?" It was thus that the splendid school at Brighton lost itself in the haze of larger questions, though the fear that it would provoke Ida to leap into the breach subsided with her prolonged, her quite shameless non-appearance. Her daughter and her successor were therefore left to gaze in united but helpless blankness at all Maisie was not learning.

This quantity was so great as to fill the child's days with a sense of intermission to which even French Lisette gave no accent—with finished games and unanswered questions and dreaded tests; with the habit, above all, in her watch for a change, of hanging over banisters when the door-bell sounded. This was the great refuge of her impatience, but what she heard at such times was a clatter of gaiety downstairs; the impression of which, from her earliest childhood, had built up in her the belief that the grown-up time was the time of real amusement and above all of real intimacy. Even Lisette, even Mrs. Wix had never, she felt, in spite of hugs and tears, been so intimate with her as so many persons at present were with Mrs. Beale and as so many others of old had been with Mrs. Farange. The note of hilarity brought people together still more than the note of melancholy, which was the one exclusively sounded, for instance, by poor

Mrs. Wix. Maisie in these days preferred none the less that domestic revels should be wafted to her from a distance: she felt sadly unsupported for facing the inquisition of the drawing-room. That was a reason the more for making the most of Susan Ash, who in her quality of under-housemaid moved at a very different level and who, none the less, was much depended upon out of doors. She was a guide to peregrinations that had little in common with those intensely definite airings that had left with the child a vivid memory of the regulated mind of Moddle. There had been under Moddle's system no dawdles at shop-windows and no nudges, in Oxford Street, of "I *say*, look at '*er*!" There had been an inexorable treatment of crossings and a serene exemption from the fear that—especially at corners, of which she was yet weakly fond—haunted the housemaid, the fear of being, as she ominously said, "spoken to." The dangers of the town equally with its diversions added to Maisie's sense of being untutored and unclaimed.

The situation, however, had taken a twist when, on another of her returns, at Susan's side, extremely tired, from the pursuit of exercise qualified by much hovering, she encountered another emotion. She on this occasion learnt at the door that her instant attendance was requested in the drawing-room. Crossing the threshold in a cloud of shame she discerned through the blur Mrs. Beale seated there with a gentleman who immediately drew the pain from her predicament by rising before her as the original of the photograph of Sir Claude. She felt the moment she looked at him that he was by far the most shining presence that had ever made her gape, and her pleasure in seeing him, in knowing that he took hold of her and kissed her, as quickly throbbed into a strange shy pride in him, a perception of his making up for her fallen state, for Susan's public nudges, which quite bruised her, and for all the lessons that, in the dead school-room, where at times she was almost afraid to stay alone, she was bored with not having. It was as if he had told her on the spot that he belonged to her, so that she could already show him off and see the effect he produced. No, nothing else that was most beautiful ever belonging to her could kindle that particular joy—not Mrs. Beale at that very moment, not papa when he was gay, nor mamma when she was dressed, nor Lisette when she was new. The joy almost overflowed in tears when he laid his hand on her and drew her to him, telling her, with a smile of which the promise was as bright as that of a Christmas-tree, that he knew her ever so well by her mother, but had come to see her now so that he might know her for himself. She could see that his view of this kind of knowledge was to make her come away with him, and, further, that it was just what he was there for and had already been some time: arranging it with Mrs. Beale and getting on with that lady in a manner evidently not at all affected by her having on the arrival of his portrait thought of him so ill. They had grown almost intimate—or had the air of it—over their discussion; and it was still further conveyed to Maisie that Mrs. Beale had made no secret, and would make yet less of one, of all that it cost to let her go. "You seem so tremendously eager," she said to the child, "that I hope you're at least clear about Sir Claude's relation to you. It does n't appear to occur to him to give you the necessary reassurance."

Maisie, a trifle mystified, turned quickly to her new friend. "Why it's of course that you're *married* to her, is n't it?"

Her anxious emphasis started them off, as she had learned to call it; this was the echo she infallibly and now quite resignedly produced; moreover Sir Claude's laughter was an indistinguishable part of the sweetness of his being there. "We've been married, my dear child, three months, and my interest in you is a consequence, don't you know? of my great affection for your mother. In coming here it's of course for your mother I'm acting."

"Oh I know,' Maisie said with all the candour of her competence. "She can't come herself—except just to the door." Then as she thought afresh: "Can't she come even to the door now?"

"There you are!" Mrs. Beale exclaimed to Sir Claude. She spoke as if his dilemma were ludicrous.

His kind face, in an hesitation, seemed to recognise it; but he answered the child with a frank smile. "No—not very well."

"Because she has married you?"

He promptly accepted this reason. "Well, that has a good deal to do with it."

He was so delightful to talk to that Maisie pursued the subject. "But papa—*he* has married Miss Overmore."

"Ah you'll see that he won't come for you at your mother's," that lady interposed.

"Yes, but that won't be for a long time," Maisie hastened to respond.

"We won't talk about it now—you've months and months to put in first." And Sir Claude drew her closer.

"Oh that's what makes it so hard to give her up!" Mrs. Beale made this point with her arms out to her stepdaughter. Maisie, quitting Sir Claude, went over to them and, clasped in a still tenderer embrace, felt entrancingly the extension of the field of happiness. "*I'll* come for you," said her stepmother, "if Sir Claude keeps you too long: we must make him quite understand that! Don't talk to me about her ladyship!" she went on to their visitor so familiarly that it was almost as if they must have met before. "I know her ladyship as if I had made her. They're a pretty pair of parents!" cried Mrs. Beale.

Maisie had so often heard them called so that the remark diverted her but an instant from the agreeable wonder of this grand new form of allusion to her mother; and that, in its turn, presently left her free to catch at the pleasant possibility, in connexion with herself, of a relation much happier as between Mrs. Beale and Sir Claude than as between mamma and papa. Still the next thing that happened was that her interest in such a relation brought to her lips a fresh question. "Have you seen papa?" she asked of Sir Claude.

It was the signal for their going off again, as her small stoicism had perfectly taken for granted that it would be. All that Mrs. Beale had nevertheless to add was the vague apparent sarcasm: "Oh papa!"

"I'm assured he's not at home," Sir Claude replied to the child; "but if he had been I should have hoped for the pleasure of seeing him."

"Won't he mind your coming?" Maisie asked as with need of the knowledge.

"Oh you bad little girl!" Mrs. Beale humorously protested.

The child could see that at this Sir Claude, though still moved to mirth, coloured a little; but he spoke to her very kindly. "That's just what I came to see, you know—whether your father *would* mind. But Mrs. Beale appears strongly of the opinion that he won't."

This lady promptly justified that view to her stepdaughter. "It will be very interesting, my dear, you know, to find out what it is to-day that your father does mind. I'm sure *I* don't know!"—and she seemed to repeat, though with perceptible resignation, her plaint of a moment before. "Your father, darling, is a very odd person indeed." She turned with this, smiling, to Sir Claude. "But perhaps it's hardly civil for me to say that of his not objecting to have *you* in the house. If you knew some of the people he does have!"

Maisie knew them all, and none indeed were to be compared to Sir Claude. He laughed back at Mrs. Beale; he looked at such moments quite as Mrs. Wix, in the long stories she told her pupil, always described the lovers of her distressed beauties—"the perfect gentleman and strikingly handsome." He got up, to the child's regret, as if he were going. "Oh I dare say we should be all right!"

Mrs. Beale once more gathered in her little charge, holding her close and looking thoughtfully over her head at their visitor. "It's so charming—for a man of your type—to have wanted her so much!"

"What do you know about my type?" Sir Claude laughed. "Whatever it may be I dare say it deceives you. The truth about me is simply that I'm the most unappreciated of—what do you call the fellows?—'family-men.' Yes, I'm a family-man; upon my honour I am!"

"Then why on earth," cried Mrs. Beale, "did n't you marry a family-woman?"

Sir Claude looked at her hard. "*You* know who one marries, I think. Besides, there *are* no family-women—hanged if there are! None of them want any children—hanged if they do!"

His account of the matter was most interesting, and Maisie, as if it were of bad omen for her, stared at the picture in some dismay. At the same time she felt, through encircling arms, her protectress hesitate. "You do come out with things! But you mean her ladyship does n't want any—really?"

"Won't hear of them—simply. But she can't help the one she *has* got." And with this Sir Claude's eyes rested on the little girl in a way that seemed to her to mask her mother's attitude with the consciousness of his own. "She must make the best of her, don't you see? If only for the look of the thing, don't you know? one wants one's wife to take the proper line about her child."

"Oh I know what one wants!" Mrs. Beale cried with a competence that evidently impressed her interlocutor.

"Well, if you keep *him* up—and I dare say you've had worry enough—why should n't I keep Ida? What's sauce for the goose is sauce for the gander—or the other way round, don't you know? I mean to see the thing through."

Mrs. Beale, for a minute, still with her eyes on him as he leaned upon the chimney piece, appeared to turn this over. "You're just a wonder of kindness—that's what you are!" she said at last. "A lady's expected to have natural feelings. But *your* horrible sex—! Is n't it a horrible sex, little love?" she demanded with her cheek upon her stepdaughter's.

"Oh I like gentlemen best," Maisie lucidly replied.

The words were taken up merrily. "That's a good one for *you*!" Sir Claude exclaimed to Mrs. Beale.

"No," said that lady: "I've only to remember the women she sees at her mother's."

"Ah they're very nice now," Sir Claude returned.

"What do you call 'nice'?"

"Well, they're all right."

"That doesn't answer me," said Mrs. Beale; "but I dare say you do take care of them. That makes you more of an angel to want this job too." And she playfully whacked her smaller companion.

"I'm not an angel—I'm an old grandmother," Sir Claude declared. "I like babies—I always did. If we go to smash I shall look for a place as responsible nurse."

Maisie, in her charmed mood, drank in an imputation on her years which at another moment might have been bitter; but the charm was sensibly interrupted by Mrs. Beale's screwing her round and gazing fondly into her eyes, "You're willing to leave me, you wretch?"

The little girl deliberated; even this consecrated tie had become as a cord she must suddenly snap. But she snapped it very gently. "Isn't it my turn for mamma?"

"You're a horrible little hypocrite! The less, I think, now said about 'turns' the better," Mrs. Beale made answer. "*I* know whose turn it is. You've not such a passion for your mother!"

"I say, I say: *do* look out!" Sir Claude quite amiably protested.

"There's nothing she hasn't heard. But it doesn't matter —it hasn't spoiled her. If you knew what it costs me to part with you!" she pursued to Maisie.

Sir Claude watched her as she charmingly clung to the child. "I'm so glad you really care for her. That's so much to the good."

Mrs. Beale slowly got up, still with her hands on Maisie, but emitting a soft exhalation. "Well, if you're glad, that may help us; for I assure you that I shall never give up any rights in her that I may consider I've acquired by my own sacrifices. I shall hold very fast to my interest in her. What seems to have happened is that she has brought you and me together."

"She has brought you and me together," said Sir Claude.

His cheerful echo prolonged the happy truth, and Maisie

broke out almost with enthusiasm: "I've brought you and her together!"

Her companions of course laughed anew and Mrs. Beale gave her an affectionate shake. "You little monster—take care what you do! But that's what she does do," she continued to Sir Claude. "She did it to me and Beale."

"Well then," he said to Maisie, "you must try the trick at *our* place." He held out his hand to her again. "Will you come now?"

"Now—just as I am?" She turned with an immense appeal to her stepmother, taking a leap over the mountain of "mending," the abyss of packing that had loomed and yawned before her. "Oh *may* I?"

Mrs. Beale addressed her assent to Sir Claude. "As well so as any other way. I'll send on her things to-morrow." Then she gave a tug to the child's coat, glancing at her up and down with some ruefulness. "She's not turned out as I should like—her mother will pull her to pieces. But what's one to do—with nothing to do it on? And she's better than when she came—you can tell her mother that. I'm sorry to have to say it to you—but the poor child was a sight."

"Oh I'll turn her out myself!" the visitor cordially said.

"I shall like to see how!"—Mrs. Beale appeared much amused. "You must bring her to show me—we can manage that. Good-bye, little fright!" And her last word to Sir Claude was that she would keep him up to the mark.

IX

THE idea of what she was to make up and the prodigious total it came to were kept well before Maisie at her mother's. These things were the constant occupation of Mrs. Wix, who arrived there by the back stairs, but in tears of joy, the day after her own arrival. The process of making up, as to which the good lady had an immense deal to say, took, through its successive phases, so long that it heralded a term at least equal to the child's last stretch with her father. This, however, was a fuller and richer time: it bounded along to the tune of Mrs. Wix's constant insistence on the energy they must both put forth. There was a fine intensity in the way the child agreed with her that under Mrs. Beale and Susan Ash she had learned nothing whatever; the wildness of the rescued castaway was one of the forces that would henceforth make for a career of conquest. The year therefore rounded itself as a receptacle of retarded knowledge—a cup brimming over with the sense that now at least she was learning. Mrs. Wix fed this sense from the stores of her conversation and with the immense bustle of her reminder that they must cull the fleeting hour. They were surrounded with subjects they must take at a rush and perpetually getting into the attitude of triumphant attack. They had certainly no idle hours, and the child went to bed each night as tired as from a long day's play. This had begun from the moment of their reunion, begun with all Mrs. Wix had to tell her young friend of the reasons of her ladyship's extraordinary behaviour at the very first.

It took the form of her ladyship's refusal for three days to see her little girl—three days during which Sir Claude made hasty merry dashes into the schoolroom to smooth down the odd situation, to say "She'll come round, you know; I assure you she'll come round," and a little even to compensate Maisie for the indignity he had caused her to suffer. There had never in the child's life been, in all ways, such a delightful amount of reparation. It came out by his sociable admission that her ladyship had not known of his visit to her late husband's house and of his having made that person's daughter a pretext for striking up an acquaintance with the dreadful creature installed there. Heaven knew she wanted her child back and had made every plan of her own for removing her; what she couldn't for the present at least forgive any one concerned was such an officious underhand way of bringing about the transfer. Maisie carried more of the weight of this resentment than even Mrs. Wix's confidential ingenuity could lighten for her, especially as Sir Claude himself was not at all ingenious, though indeed on the other hand he was not at all crushed. He was amused and intermittent and at moments most startling; he impressed on his young companion, with a frankness that agitated her much more than he seemed to guess, that he depended on her not letting her mother, when she should see her, get anything out of her about anything Mrs. Beale might have said to him. He came in and out; he professed, in joke, to take tremendous precautions; he showed a positive disposition to romp. He chaffed Mrs. Wix till she was purple with the pleasure of it, and reminded Maisie of the reticence he expected of her till she set her teeth like an Indian captive. Her lessons these first days and indeed for long after seemed to be all about Sir Claude, and yet she never really mentioned to Mrs. Wix that she was prepared, under his inspiring injunction, to be vainly tortured. This lady, however, had formulated the position of things with an acuteness that showed how little she needed to be coached. Her explanation of everything that seemed not quite pleasant—and if her own footing was perilous it met that danger as well—was that her ladyship was passionately in love. Maisie accepted this hint with infinite awe and pressed upon it much when she was at last summoned into the presence of her mother.

There she encountered matters amid which it seemed really to help to give her a clue—an almost terrifying strangeness, full, none the less, after a little, of reverberations of Ida's old fierce and demonstrative recoveries of possession. They had been some time in the house together, and this demonstration came late. Preoccupied, however, as Maisie was with the idea of the sentiment Sir Claude had inspired, and familiar, in addition, by Mrs. Wix's anecdotes, with the ravages that in general such a sentiment could produce, she was able to make allowances for her ladyship's remarkable appearance, her violent splendour, the wonderful colour of her lips and even the hard stare, the stare of some gorgeous idol described in a story-book, that had come into her eyes in consequence of a curious thickening of their already rich circumference. Her professions and explanations were mixed with eager challenges and sudden drops, in the midst of which Maisie recognised as a memory of other years the rattle of her trinkets and the scratch of her endearments, the odour of her clothes and the jumps of her conversation. She had all her old clever way—Mrs. Wix said it was "aristocratic"—of changing the subject as she might have slammed the door in your face. The principal thing that was different was the tint of her golden hair, which had changed to a coppery red and, with the head it profusely covered, struck the child as now lifted still further aloft. This picturesque parent showed literally a grander stature and a nobler presence, things which, with some others that might have been bewildering, were handsomely accounted for by the romantic state of her affections. It was her affections, Maisie could easily see, that led Ida to break out into questions as to what had passed at the other house between that horrible woman and Sir Claude; but it was also just here that the little girl was able to recall the effect with which in earlier days she had practised the pacific art of stupidity. This art again came to her aid: her mother, in getting rid of her after an interview in which she had achieved a hollowness beyond her years, allowed her fully to understand she had not grown a bit more amusing.

She could bear that; she could bear anything that helped her to feel she had done something for Sir Claude. If she had n't told Mrs. Wix how Mrs. Beale seemed to like him she certainly could n't tell her ladyship. In the way the past revived for her there was a queer confusion. It was because mamma hated papa that she used to want to know bad things of him; but if at present she wanted to know the same of Sir Claude it was quite from the opposite motive. She was awestruck at the manner in which a lady might be affected through the passion mentioned by Mrs. Wix; she held her breath with the sense of picking her steps among the tremendous things of life. What she did, however, now, after the interview with her mother, impart to Mrs. Wix was that, in spite of her having had her "good" effect, as she

with her stepfather, though she had had little to say about it to Mrs. Wix, she had during the first weeks of her stay at her mother's found more than one opportunity to revert. As to what had taken place the day Sir Claude came for her, she had been vaguely grateful to Mrs. Wix for not attempting, as her mother had attempted, to put her through. That was what Sir Claude had called the process when he warned her of it, and again afterwards when he told her she was an awfully good "chap" for having foiled it. Then it was that, well aware Mrs. Beale had n't in the least really given her up, she had asked him if he remained in communication with her and if for the time everything must really be held to be at an end between her stepmother and herself. This conversation had occurred in consequence of his one day popping into the schoolroom and finding Maisie alone.

X

HE was smoking a cigarette and he stood before the fire and looked at the meagre appointments of the room in a way that made her rather ashamed of them. Then before (on the subject of Mrs. Beale) he let her "draw" him—that was another of his words; it was astonishing how many she gathered in— he remarked that really mamma kept them rather low on the question of decorations. Mrs. Wix had put up a Japanese fan and two rather grim texts; she had wished they were gayer, but they were all she happened to have. Without Sir Claude's photograph, however, the place would have been, as he said, as dull as a cold dinner. He had said as well that there were all sorts of things they ought to have; yet governess and pupil, it had to be admitted, were still divided between discussing the places where any sort of thing would look best if any sort of thing should ever come and acknowledging that mutability in the child's career which was naturally unfavourable to accumulation. She stayed long enough only to miss things, not half long enough to deserve them. The way Sir Claude looked about the schoolroom had made her feel with humility as if it were not very different from the shabby attic in which she had visited Susan Ash. Then he had said in abrupt reference to Mrs. Beale: "Do you think she really cares for you?"

"Oh awfully!" Maisie had replied.

"But, I mean, does she love you for yourself, as they call it, don't you know? Is she as fond of you, now, as Mrs. Wix?"

The child turned it over. "Oh I'm not every bit Mrs. Beale has!"

Sir Claude seemed much amused at this. "No; you're not every bit she has!"

He laughed for some moments, but that was an old story to Maisie, who was not too much disconcerted to go on: "But she'll never give me up."

"Well, I won't either, old boy: so that's not so wonderful, and she's not the only one. But if she's so fond of you, why doesn't she write to you?"

"Oh on account of mamma." This was rudimentary, and she was almost surprised at the simplicity of Sir Claude's question.

"I see—that's quite right," he answered. "She might get at you—there are all sorts of ways. But of course there's Mrs. Wix."

"There's Mrs. Wix," Maisie lucidly concurred. "Mrs. Wix can't abide her."

Sir Claude seemed interested. "Oh she can't abide her? Then what does she say about her?"

"Nothing at all—because she knows I shouldn't like it. Isn't it sweet of her?" the child asked.

"Certainly; rather nice. Mrs. Beale wouldn't hold her tongue for any such thing as that, would she?"

Maisie remembered how little she had done so; but she desired to protect Mrs. Beale too. The only protection she could think of, however, was the plea: "Oh at papa's, you know, they don't mind!"

At this Sir Claude only smiled. "No, I dare say not. But

here we mind, don't we?—we take care what we say. I don't suppose it's a matter on which I ought to prejudice you," he went on; "but I think we must on the whole be rather nicer here than at your father's. However, I don't press that; for it's the sort of question on which it's awfully awkward for you to speak. Don't worry, at any rate: I assure you I'll back you up." Then after a moment and while he smoked he reverted to Mrs. Beale and the child's first enquiry. "I'm afraid we can't do much for her just now. I haven't seen her since that day—upon my word I haven't seen her." The next instant, with a laugh the least bit foolish, the young man slightly coloured: he must have felt this profession of innocence to be excessive as addressed to Maisie. It was inevitable to say to her, however, that of course her mother loathed the lady of the other house. He couldn't go there again with his wife's consent, and he wasn't the man—he begged her to believe, falling once more, in spite of himself, into the scruple of showing the child he didn't trip—to go there without it. He was liable in talking with her to take the tone of her being also a man of the world. He had gone to Mrs. Beale's to fetch away Maisie, but that was altogether different. Now that she was in her mother's house what pretext had he to give her mother for paying calls on her father's wife? And of course Mrs. Beale couldn't come to Ida's—Ida would tear her limb from limb. Maisie, with this talk of pretexts, remembered how much Mrs. Beale had made of her being a good one, and how, for such a function, it was her fate to be either much depended on or much missed. Sir Claude moreover recognised on this occasion that perhaps things would take a turn later on; and he wound up by saying: "I'm sure she does sincerely care for you— how can she possibly help it? She's very young and very pretty and very clever: I think she's charming. But we must walk very straight. If you'll help me, you know, I'll help *you*," he concluded in the pleasant fraternising, equalising, not a bit patronising way which made the child ready to go through anything for him and the beauty of which, as she dimly felt, was that it was so much less a deceitful descent to her years than a real indifference to them.

It gave her moments of secret rapture—moments of believing she might help him indeed. The only mystification in this was the imposing time of life that her elders spoke of as youth. For Sir Claude then Mrs. Beale was "young," just as for Mrs. Wix Sir Claude was: that was one of the merits for which Mrs. Wix most commended him. What therefore was Maisie herself, and, in another relation to the matter, what therefore was mamma? It took her some time to puzzle out with the aid of an experiment or two that it would n't do to talk about mamma's youth. She even went so far one day, in the presence of that lady's thick colour and marked lines, as to wonder if it would occur to any one but herself to do so. Yet if she was n't young then she was old; and this threw an odd light on her having a husband of a different generation. Mr. Farange was still older—that Maisie perfectly knew; and it brought her in due course to the perception of how much more, since Mrs. Beale was younger than Sir Claude, papa must be older than Mrs. Beale. Such discoveries were disconcerting and even a trifle confounding: these persons, it appeared, were not of the age they ought to be. This was somehow particularly the case with mamma, and the fact made her reflect with some relief on her not having gone· with Mrs. Wix into the question of Sir Claude's attachment to his wife. She was conscious that in confining their attention to the state of her ladyship's own affections they had been controlled—Mrs. Wix perhaps in especial—by delicacy and even by embarrassment. The end of her colloquy with her stepfather in the schoolroom was her saying: "Then if we 're not to see Mrs. Beale at all it is n't what she seemed to think when you came for me."

He looked rather blank. "What did she seem to think?"

"Why that I 've brought you together."

"She thought that?" Sir Claude asked.

Maisie was surprised at his already forgetting it. "Just

as I had brought papa and her. Don't you remember she said so?"

It came back to Sir Claude in a peal of laughter. "Oh yes—she said so!"

"And *you* said so," Maisie lucidly pursued.

He recovered, with increasing mirth, the whole occasion. "And *you* said so!" he retorted as if they were playing a game.

"Then were we all mistaken?"

He considered a little. "No, on the whole not. I dare say it's just what you *have* done. We *are* together—it's really most odd. She's thinking of us—of you and me—though we don't meet. And I've no doubt you'll find it will be all right when you go back to her."

"Am I going back to her?" Maisie brought out with a little gasp which was like a sudden clutch of the happy present.

It appeared to make Sir Claude grave a moment; it might have made him feel the weight of the pledge his action had given. "Oh some day, I suppose! We've plenty of time."

"I've such a tremendous lot to make up," Maisie said with a sense of great boldness.

"Certainly, and you must make up every hour of it. Oh I'll *see* that you do!"

This was encouraging; and to show cheerfully that she was reassured she replied: "That's what Mrs. Wix sees too."

"Oh yes," said Sir Claude; "Mrs. Wix and I are shoulder to shoulder."

Maisie took in a little this strong image; after which she exclaimed: "Then I've done it also to you and her—I've brought *you* together!"

"Blest if you have n't!" Sir Claude laughed. "And more, upon my word, than any of the lot. Oh you've done for *us*! Now if you could—as I suggested, you know, that day—only manage me and your mother!"

The child wondered. "Bring you and *her* together?"

"You see we're not together—not a bit. But I ought n't to tell you such things; all the more that you won't really do it—not you. No, old chap," the young man continued;

"there you'll break down. But it won't matter—we'll rub along. The great thing is that you and I are all right."

"*We're* all right!" Maisie echoed devoutly. But the next moment, in the light of what he had just said, she asked: "How shall I ever leave you?" It was as if she must somehow take care of him.

His smile did justice to her anxiety. "Oh well, you need n't! It won't come to that."

"Do you mean that when I do go you'll go with me?"

Sir Claude cast about. "Not exactly 'with' you perhaps; but I shall never be far off."

"But how do you know where mamma may take you?"

He laughed again. "I don't, I confess!" Then he had an idea, though something too jocose. "That will be for you to see—that she shan't take me too far."

"How can I help it?" Maisie enquired in surprise. "Mamma does n't care for me," she said very simply. "Not really." Child as she was, her little long history was in the words; and it was as impossible to contradict her as if she had been venerable.

Sir Claude's silence was an admission of this, and still more the tone in which he presently replied: "That won't prevent her from—some time or other—leaving me with you."

"Then we'll live together?" she eagerly demanded.

"I'm afraid," said Sir Claude, smiling, "that that will be Mrs. Beale's real chance."

Her eagerness just slightly dropped at this; she remembered Mrs. Wix's pronouncement that it was all an extraordinary muddle. "To take me again? Well, can't you come to see me there?"

"Oh I dare say!"

Though there were parts of childhood Maisie had lost she had all childhood's preference for the particular promise. "Then you *will* come—you'll come often, won't you?" she insisted; while at the moment she spoke the door opened for the return of Mrs. Wix. Sir Claude hereupon, instead of replying, gave her a look which left her silent and embarrassed.

When he again found privacy convenient, however—which happened to be long in coming—he took up their conversation very much where it had dropped. "You see, my dear, if I shall be able to go to you at your father's it yet is n't at all the same thing for Mrs. Beale to come to you here." Maisie gave a thoughtful assent to this proposition, though conscious she could scarcely herself say just where the difference would lie. She felt how much her stepfather saved her, as he said with his habitual amusement, the trouble of that. "I shall probably be able to go to Mrs. Beale's without your mother's knowing it."

Maisie stared with a certain thrill at the dramatic element in this. "And she could n't come here without mamma's—?" She was unable to articulate the word for what mamma would do.

"My dear child, Mrs. Wix would tell of it."

"But I thought," Maisie objected, "that Mrs. Wix and you—"

"Are such brothers-in-arms?"—Sir Claude caught her up. "Oh yes, about everything but Mrs. Beale. And if you should suggest," he went on, "that we might somehow or other hide her peeping in from Mrs. Wix—"

"Oh I don't suggest *that*!" Maisie in turn cut him short.

Sir Claude looked as if he could indeed quite see why. "No; it would really be impossible." There came to her from this glance at what they might hide the first small glimpse of something in him that she would n't have expected. There had been times when she had had to make the best of the impression that she was herself deceitful; yet she had never concealed anything bigger than a thought. Of course she now concealed this thought of how strange it would be to see *him* hide; and while she was so actively engaged he continued: "Besides, you know, I'm not afraid of your father."

"And you are of my mother?"

"Rather, old man!" Sir Claude returned.

XI

IT must not be supposed that her ladyship's intermissions
were not qualified by demonstrations of another order—
triumphal entries and breathless pauses during which she
seemed to take of everything in the room, from the state of
the ceiling to that of her daughter's boot-toes, a survey that
was rich in intentions. Sometimes she sat down and sometimes she surged about, but her attitude wore equally in
either case the grand air of the practical. She found so much
to deplore that she left a great deal to expect, and bristled
so with calculation that she seemed to scatter remedies and
pledges. Her visits were as good as an outfit; her manner, as
Mrs. Wix once said, as good as a pair of curtains; but she
was a person addicted to extremes—sometimes barely speaking to her child and sometimes pressing this tender shoot to
a bosom cut, as Mrs. Wix had also observed, remarkably
low. She was always in a fearful hurry, and the lower the
bosom was cut the more it was to be gathered she was
wanted elsewhere. She usually broke in alone, but sometimes
Sir Claude was with her, and during all the earlier period
there was nothing on which these appearances had had so
delightful a bearing as on the way her ladyship was, as Mrs.
Wix expressed it, under the spell. "But is n't she under it!"
Maisie used in thoughtful but familiar reference to exclaim
after Sir Claude had swept mamma away in peals of natural
laughter. Not even in the old days of the convulsed ladies
had she heard mamma laugh so freely as in these moments
of conjugal surrender, to the gaiety of which even a little girl
could see she had at last a right—a little girl whose thought-
fulness was now all happy selfish meditation on good omens
and future fun.

Unaccompanied, in subsequent hours, and with an effect
of changing to meet a change, Ida took a tone superficially
disconcerting and abrupt—the tone of having, at an immense cost, made over everything to Sir Claude and wishing

others to know that if everything was n't right it was because Sir Claude was so dreadfully vague. "He has made from the first such a row about you," she said on one occasion to Maisie, "that I've told him to do for you himself and try how he likes it—see? I've washed my hands of you; I've made you over to him; and if you're discontented it's on him, please, you'll come down. So don't haul poor *me* up— I assure you I've worries enough." One of these, visibly, was that the spell rejoiced in by the schoolroom fire was already in danger of breaking; another was that she was finally forced to make no secret of her husband's unfitness for real responsibilities. The day came indeed when her breathless auditors learnt from her in bewilderment that what ailed him was that he was, alas, simply not serious. Maisie wept on Mrs. Wix's bosom after hearing that Sir Claude was a butterfly; considering moreover that her governess but half-patched it up in coming out at various moments the next few days with the opinion that it was proper to his "station" to be careless and free. That had been proper to every one's station that she had yet encountered save poor Mrs. Wix's own, and the particular merit of Sir Claude had seemed precisely· that he was different from every one. She talked with him, however, as time went on, very freely about her mother; being with him, in this relation, wholly without the fear that had kept her silent before her father—the fear of bearing tales and making bad things worse. He appeared to accept the idea that he had taken her over and made her, as he said, his particular lark; he quite agreed also that he was an awful fraud and an idle beast and a sorry dunce. And he never said a word to her against her mother—he only remained dumb and discouraged in the face of her ladyship's own overtopping earnestness. There were occasions when he even spoke as if he had wrenched his little charge from the arms of a parent who had fought for her tooth and nail.

This was the very moral of a scene that flashed into vividness one day when the four happened to meet without company in the drawing-room and Maisie found herself clutched to her mother's breast and passionately sobbed and shrieked over, made the subject of a demonstration evidently sequent to some sharp passage just enacted. The connexion required that while she almost cradled the child in her arms Ida should speak of her as hideously, as fatally estranged, and should rail at Sir Claude as the cruel author of the outrage. "He has taken you *from* me," she cried; "he has set you *against* me, and you've been won away and your horrid little mind has been poisoned! You've gone over to him, you've given yourself up to side against me and hate me. You never open your mouth to me—you know you don't; and you chatter to him like a dozen magpies. Don't lie about it—I hear you all over the place. You hang about him in a way that's barely decent—he can do what he likes with you. Well then, let him, to his heart's content: he has been in such a hurry to take you that we'll see if it suits him to keep you. I'm very good to break my heart about it when you've no more feeling for me than a clammy little fish!" She suddenly thrust the child away and, as a disgusted admission of failure, sent her flying across the room into the arms of Mrs. Wix, whom at this moment and even in the whirl of her transit Maisie saw, very red, exchange a quick queer look with Sir Claude.

The impression of the look remained with her, confronting her with such a critical little view of her mother's explosion that she felt the less ashamed of herself for incurring the reproach with which she had been cast off. Her father had once called her a heartless little beast, and now, though decidedly scared, she was as stiff and cold as if the description had been just. She was not even frightened enough to cry, which would have been a tribute to her mother's wrongs: she was only, more than anything else, curious about the opinion mutely expressed by their companions. Taking the earliest opportunity to question Mrs. Wix on this subject she elicited the remarkable reply: "Well, my dear, it's her ladyship's game, and we must just hold on like grim death." Maisie could interpret at her leisure these ominous words. Her reflexions indeed at this moment thickened apace, and one of them made her sure that her governess had conversations, private, earnest and not infrequent, with her denounced stepfather. She perceived in the light of a second episode that something beyond her knowledge had taken place in the house. The things beyond her knowledge—numerous enough in truth—had not hitherto, she believed, been the things that had been nearest to her: she had even had in the past a small smug conviction that in the domestic labyrinth she always kept the clue. This time too, however, she at last found out—with the discreet aid, it had to be confessed, of Mrs. Wix. Sir Claude's own assistance was abruptly taken from her, for his comment on her ladyship's game was to start on the spot, quite alone, for Paris, evidently because he wished to show a spirit when accused of bad behaviour. He might be fond of his stepdaughter, Maisie felt, without wishing her to be after all thrust on him in such a way; his absence therefore, it was clear, was a protest against the thrusting. It was while this absence lasted that our young lady finally discovered what had happened in the house to be that her mother was no longer in love.

The limit of a passion for Sir Claude had certainly been reached, she judged, some time before the day on which her ladyship burst suddenly into the schoolroom to introduce Mr. Perriam, who, as she announced from the doorway to Maisie, would n't believe his ears that one had a great hoyden of a daughter. Mr. Perriam was short and massive—Mrs. Wix remarked afterwards that he was "too fat for the pace"; and it would have been difficult to say of him whether his head were more bald or his black moustache more bushy. He seemed also to have moustaches over his eyes, which, however, by no means prevented these polished little globes from rolling round the room as if they had been billiard-balls impelled by Ida's celebrated stroke. Mr. Perriam wore on the hand that pulled his moustache a diamond of dazzling lustre, in consequence of which and of his general weight and mystery our young lady observed on his departure that if he had only had a turban he would have been quite her idea of a heathen Turk.

"He's quite my idea," Mrs. Wix replied, "of a heathen Jew."

"Well, I mean," said Maisie, "of a person who comes from the East."

"That's where he *must* come from," her governess opined—"he comes from the City." In a moment she added as if she knew all about him. "He's one of those people who have lately broken out. He'll be immensely rich."

"On the death of his papa?" the child interestedly enquired.

"Dear no—nothing hereditary. I mean he has made a mass of money."

"How much, do you think?" Maisie demanded.

Mrs. Wix reflected and sketched it. "Oh many millions."

"A hundred?"

Mrs. Wix was not sure of the number, but there were enough of them to have seemed to warm up for the time the penury of the schoolroom—to linger there as an afterglow of the hot heavy light Mr. Perriam sensibly shed. This was also, no doubt, on his part, an effect of that enjoyment of life with which, among her elders, Maisie had been in contact from her earliest years—the sign of happy maturity, the old familiar note of overflowing cheer. "How d'ye do, ma'am? How d'ye do, little miss?"—he laughed and nodded at the gaping figures. "She has brought me up for a peep—it's true I would n't take you on trust. She's always talking about you, but she'd never produce you; so to-day I challenged her on the spot. Well, you ain't a myth, my dear—I back down on that," the visitor went on to Maisie; "nor you either, miss, though you might be, to be sure!"

"I bored him with you, darling—I bore every one," Ida said, "and to prove that you *are* a sweet thing, as well as a fearfully old one, I told him he could judge for himself. So now he sees that you're a dreadful bouncing business and that your poor old Mummy's at least sixty!"—and her ladyship smiled at Mr. Perriam with the charm that her daughter had heard imputed to her at papa's by the merry gentlemen

who had so often wished to get from him what they called a
"rise." Her manner at that instant gave the child a glimpse
more vivid than any yet enjoyed of the attraction that
papa, in remarkable language, always denied she could put
forth.

Mr. Perriam, however, clearly recognised it in the humour
with which he met her. "I never said you ain't wonderful—
did I ever say it, hey?" and he appealed with pleasant confidence to the testimony of the schoolroom, about which
itself also he evidently felt something might be expected of
him. "So this is their little place, hey? Charming, charming,
charming!" he repeated as he vaguely looked round. The
interrupted students clung together as if they had been
personally exposed; but Ida relieved their embarrassment
by a hunch of her high shoulders. This time the smile she
addressed to Mr. Perriam had a beauty of sudden sadness.
"What on earth is a poor woman to do?"

The visitor's grimace grew more marked as he continued
to look, and the conscious little schoolroom felt still more like
a cage at a menagerie. "Charming, charming, charming!"
Mr. Perriam insisted; but the parenthesis closed with a
prompt click. "There you are!" said her ladyship. "By-bye!"
she sharply added. The next minute they were on the stairs,
and Mrs. Wix and her companion, at the open door and looking mutely at each other, were reached by the sound of the
large social current that carried them back to their life.

It was singular perhaps after this that Maisie never put a
question about Mr. Perriam, and it was still more singular
that by the end of a week she knew all she did n't ask. What
she most particularly knew—and the information came to
her, unsought, straight from Mrs. Wix—was that Sir Claude
would n't at all care for the visits of a millionaire who was in
and out of the upper rooms. How little he would care was
proved by the fact that under the sense of them Mrs. Wix's
discretion broke down altogether; she was capable of a
transfer of allegiance, capable, at the altar of propriety, of a
desperate sacrifice of her ladyship. As against Mrs. Beale,
she more than once intimated, she had been willing to do

the best for her, but as against Sir Claude she could do nothing for her at all. It was extraordinary the number of things that, still without a question, Maisie knew by the time her stepfather came back from Paris—came bringing her a splendid apparatus for painting in water-colours and bringing Mrs. Wix, by a lapse of memory that would have been droll if it had not been a trifle disconcerting, a second and even a more elegant umbrella. He had forgotten all about the first, with which, buried in as many wrappers as a mummy of the Pharaohs, she would n't for the world have done anything so profane as use it. Maisie knew above all that though she was now, by what she called an informal understanding, on Sir Claude's "side," she had yet not uttered a word to him about Mr. Perriam. That gentleman became therefore a kind of flourishing public secret, out of the depths of which governess and pupil looked at each other portentously from the time their friend was restored to them. He was restored in great abundance, and it was marked that, though he appeared to have felt the need to take a stand against the risk of being too roughly saddled with the offspring of others, he at this period exposed himself more than ever before to the presumption of having created expectations.

If it had become now, for that matter, a question of sides, there was at least a certain amount of evidence as to where they all were. Maisie of course, in such a delicate position, was on nobody's; but Sir Claude had all the air of being on hers. If therefore Mrs. Wix was on Sir Claude's, her ladyship on Mr. Perriam's and Mr. Perriam presumably on her ladyship's, this left only Mrs. Beale and Mr. Farange to account for. Mrs. Beale clearly was, like Sir Claude, on Maisie's, and papa, it was to be supposed, on Mrs. Beale's. Here indeed was a slight ambiguity, as papa's being on Mrs. Beale's did n't somehow seem to place him quite on his daughter's. It sounded, as this young lady thought it over, very much like puss-in-the-corner, and she could only wonder if the distribution of parties would lead to a rushing to and fro and a changing of places. She was in the presence, she felt, of restless change: was n't it restless enough that her mother and her stepfather should already be on different sides? That was the great thing that had domestically happened. Mrs. Wix, besides, had turned another face: she had never been exactly gay, but her gravity was now an attitude as public as a posted placard. She seemed to sit in her new dress and brood over her lost delicacy, which had become almost as doleful a memory as that of poor Clara Matilda. "It *is* hard for him," she often said to her companion; and it was surprising how competent on this point Maisie was conscious of being to agree with her. Hard as it was, however, Sir Claude had never shown to greater advantage than in the gallant generous sociable way he carried it off: a way that drew from Mrs. Wix a hundred expressions of relief at his not having suffered it to embitter him. It threw him more and more at last into the schoolroom, where he had plainly begun to recognise that if he was to have the credit of perverting the innocent child he might also at least have the amusement. He never came into the place without telling its occupants that they were the nicest people in the house—a remark which always led them to say to each other "Mr. Perriam!" as loud as ever compressed lips and enlarged eyes could make them articulate. He caused Maisie to remember what she had said to Mrs. Beale about his having the nature of a good nurse, and, rather more than she intended before Mrs. Wix, to bring the whole thing out by once remarking to him that none of her good nurses had smoked quite so much in the nursery. This had no more effect than it was meant to on his cigarettes: he was always smoking, but always declaring that it was death to him not to lead a domestic life.

He led one after all in the schoolroom, and there were hours of late evening, when she had gone to bed, that Maisie knew he sat there talking with Mrs. Wix of how to meet his difficulties. His consideration for this unfortunate woman even in the midst of them continued to show him as the perfect gentleman and lifted the subject of his courtesy into an upper air of beatitude in which her very pride had the hush of anxiety. "He leans on me—he leans on me!" she only announced from time to time; and she was more surprised than amused when, later on, she accidentally found she had given her pupil the impression of a support literally supplied by her person. This glimpse of a misconception led her to be explicit—to put before the child, with an air of mourning indeed for such a stoop to the common, that what they talked about in the small hours, as they said, was the question of his taking right hold of life. The life she wanted him to take right hold of was the public: "she" being, I hasten to add, in this connexion, not the mistress of his fate, but only Mrs. Wix herself. She had phrases about him that were full of easy understanding, yet full of morality. "He's a wonderful nature, but he can't live like the lilies. He's all right, you know, but he must have a high interest." She had more than once remarked that his affairs were sadly involved, but that they must get him—Maisie and she together apparently—into Parliament. The child took it from her with a flutter of importance that Parliament was his natural sphere, and she was the less prepared to recognise a hindrance as she had never heard of any affairs whatever that were not involved. She had in the old days once been told by Mrs. Beale that her very own were, and with the refreshment of knowing that she *had* affairs the information had n't in the least overwhelmed her. It was true and perhaps a little alarming that she had never heard of any such matters since then. Full of charm at any rate was the prospect of some day getting Sir Claude in; especially after Mrs. Wix, as the fruit of more midnight colloquies, once went so far as to observe that she really believed it was all that was wanted to save him. This critic, with these words, struck her disciple as cropping up, after the manner of mamma when mamma talked, quite in a new place. The child stared as at the jump of a kangaroo.

"Save him from what?"

Mrs. Wix debated, then covered a still greater distance. "Why just from awful misery."

XII

SHE had not at the moment explained her ominous speech, but the light of remarkable events soon enabled her companion to read it. It may indeed be said that these days brought on a high quickening of Maisie's direct perceptions, of her sense of freedom to make out things for herself. This was helped by an emotion intrinsically far from sweet—the increase of the alarm that had most haunted her meditations. She had no need to be told, as on the morrow of the revelation of Sir Claude's danger she was told by Mrs. Wix, that her mother wanted more and more to know why the devil her father did n't send for her: she had too long expected mamma's curiosity on this point to express itself sharply. Maisie could meet such pressure so far as meeting it was to be in a position to reply, in words directly inspired, that papa would be hanged before he'd again be saddled with her. She therefore recognised the hour that in troubled glimpses she had long foreseen, the hour when—the phrase for it came back to her from Mrs. Beale—with two fathers, two mothers and two homes, six protections in all, she should n't know "wherever" to go. Such apprehension as she felt on this score was not diminished by the fact that Mrs. Wix herself was suddenly white with terror: a circumstance leading Maisie to the further knowledge that this lady was still more scared on her own behalf than on that of her pupil. A governess who had only one frock was not likely to have either two fathers or two mothers: accordingly if even with these resources Maisie was to be in the streets, where in the name of all that was dreadful was poor Mrs. Wix to be? She had had, it appeared, a tremendous brush with Ida, which had begun and ended with the request that she would be pleased on the spot to "bundle." It had come suddenly but completely, this signal of which she had gone in fear. The companions confessed to each other the dread each had hidden the worst of, but Mrs. Wix was better off than Maisie in having a plan of defence. She declined indeed to communicate it till it was quite mature; but meanwhile, she hastened to declare, her feet were firm in the schoolroom. They could only be loosened by force: she would "leave" for the police perhaps, but she would n't leave for mere outrage. That would be to play her ladyship's game, and it would take another turn of the screw to make her desert her darling. Her ladyship had come down with extraordinary violence: it had been one of many symptoms of a situation strained—"between them all," as Mrs. Wix said, "but especially between the two"—to the point of God only knew what.

Her description of the crisis made the child balance. "Between which two?—papa and mamma?"

"Dear no. I mean between your mother and *him*."

Maisie, in this, recognised an opportunity to be really deep. "'Him'?—Mr. Perriam?"

She fairly brought a blush to the scared face. "Well, my dear, I must say what you *don't* know ain't worth mentioning. That it won't go on for ever with Mr. Perriam—since I *must* meet you—who can suppose? But I meant dear Sir Claude."

Maisie stood corrected rather than abashed. "I see. But it's about Mr. Perriam he's angry?"

Mrs. Wix waited. "He says he's not."

"Not angry? He has told you so?"

Mrs. Wix looked at her hard. "Not about *him*."

"Then about some one else?"

Mrs. Wix looked at her harder. "About some one else."

"Lord Eric?" the child promptly brought forth.

At this, of a sudden, her governess was more agitated. "Oh why, little unfortunàte, should we discuss their dreadful names?"—and she threw herself for the millionth time on Maisie's neck. It took her pupil but a moment to feel that she quivered with insecurity, and, the contact of her terror aiding, the pair in another instant were sobbing in each other's arms. Then it was that, completely relaxed, demoralised as she had never been, Mrs. Wix suffered her wound to bleed and her resentment to gush. Her great bitterness was that Ida had called her false, denounced her hypocrisy and duplicity, reviled her spying and tattling, her lying and grovelling to Sir Claude. "Me, *me*," the poor woman wailed, "who've seen what I've seen and gone through everything only to cover her up and ease her off and smooth her down? If I've been an 'ipocrite it's the other way round: I've pretended, to him and to her, to myself and to you and to every one, *not* to see! It serves me right to have held my tongue before such horrors!" What horrors they were her companion forbore too closely to enquire, showing even signs not a few of an ability to take them for granted. That put the couple more than ever, in this troubled sea, in the same boat, so that with the consciousness of ideas on the part of her fellow mariner Maisie could sit close and wait. Sir Claude on the morrow came in to tea, and then the ideas were produced. It was extraordinary how the child's presence drew out their full strength. The principal one was startling, but Maisie appreciated the courage with which her governess handled it. It simply consisted of the proposal that whenever and wherever they should seek refuge Sir Claude should consent to share their asylum. On this protesting with all the warmth in nature against this note of secession she asked what else in the world was left to them if her ladyship should stop supplies.

"Supplies be hanged, my dear woman!" said their delightful friend. "Leave supplies to me—I'll take care of supplies."

Mrs. Wix rose to it. "Well, it's exactly because I knew you'd be so glad to do so that I put the question before you. There's a way to look after us better than any other. The way's just to come along with us."

It hung before Maisie, Mrs. Wix's way, like a glittering picture, and she clasped her hands in ecstasy. "Come along, come along, come along!"

Sir Claude looked from his stepdaughter back to her governess. "Do you mean leave this house and take up my abode with you?"

"It will be the right thing—if you feel as you've told me

you feel." Mrs. Wix, sustained and uplifted, was now as clear as a bell.

Sir Claude had the air of trying to recall what he had told her; then the light broke that was always breaking to make his face more pleasant. "It's your happy thought that I shall take a house for you?"

"For the wretched homeless child. Any roof—over *our* heads—will do for us; but of course for you it will have to be something really nice."

Sir Claude's eyes reverted to Maisie, rather hard, as she thought; and there was a shade in his very smile that seemed to show her—though she also felt it did n't show Mrs. Wix —that the accommodation prescribed must loom to him pretty large. The next moment, however, he laughed gaily enough. "My dear lady, you exaggerate tremendously *my* poor little needs." Mrs. Wix had once mentioned to her young friend that when Sir Claude called her his dear lady he could do anything with her; and Maisie felt a certain anxiety to see what he would do now. Well, he only addressed her a remark of which the child herself was aware of feeling the force. "Your plan appeals to me immensely; but of course—don't you see?—I shall have to consider the position I put myself in by leaving my wife."

"You'll also have to remember," Mrs. Wix replied, "that if you don't look out your wife won't give you time to consider. Her ladyship will leave *you*."

"Ah my good friend, I do look out!" the young man returned while Maisie helped herself afresh to bread and butter. "Of course if that happens I shall have somehow to turn round; but I hope with all my heart it won't. I beg your pardon," he continued to his stepdaughter, "for appearing to discuss that sort of possibility under your sharp little nose. But the fact is I *forget* half the time that Ida's your sainted mother."

"So do I!" said Maisie, her mouth full of bread and butter and to put him the more in the right.

Her protectress, at this, was upon her again. "The little desolate precious pet!" For the rest of the conversation she was enclosed in Mrs. Wix's arms, and as they sat there interlocked Sir Claude, before them with his tea-cup, looked down at them in deepening thought. Shrink together as they might they could n't help, Maisie felt, being a very large lumpish image of what Mrs. Wix required of his slim fineness. She knew moreover that this lady did n't make it better by adding in a moment: "Of course *we* should n't dream of a whole house. Any sort of little lodging, however humble, would be only too blest."

"But it would have to be something that would hold us all," said Sir Claude.

"Oh yes," Mrs. Wix concurred; "the whole point's our being together. While you're waiting, before you act, for her ladyship to take some step, our position here will come to an impossible pass. You don't know what I went through with her for you yesterday—and for our poor darling; but it's not a thing I can promise you often to face again. She cast me out in horrible language—she has instructed the servants not to wait on me."

"Oh the poor servants are all right!" Sir Claude eagerly cried.

"They're certainly better than their mistress. It's too dreadful that I should sit here and say of your wife, Sir Claude, and of Maisie's own mother, that she's lower than a domestic; but my being betrayed into such remarks is just a reason the more for our getting away. I shall stay till I'm taken by the shoulders, but that may happen any day. What also may perfectly happen, you must permit me to repeat, is that she'll go off to get rid of us."

"Oh if she'll only do that!" Sir Claude laughed. "That would be the very making of us!"

"Don't say it—don't say it!" Mrs. Wix pleaded. "Don't speak of anything so fatal. You know what I mean. We must all cling to the right. You must n't be bad."

Sir Claude set down his tea-cup; he had become more grave and he pensively wiped his moustache. "Won't all the world say I'm awful if I leave the house before—before she has bolted? They'll say it was my doing so that made her bolt."

Maisie could grasp the force of this reasoning, but it offered no check to Mrs. Wix. "Why need you mind that—if you've done it for so high a motive? Think of the beauty of it," the good lady pressed.

"Of bolting with *you*?" Sir Claude ejaculated.

She faintly smiled—she even faintly coloured. "So far from doing you harm it will do you the highest good. Sir Claude, if you'll listen to me, it will save you."

"Save me from what?"

Maisie, at this question, waited with renewed suspense for an answer that would bring the thing to some finer point than their companion had brought it to before. But there was on the contrary only more mystification in Mrs. Wix's reply. "Ah from you know what!"

"Do you mean from some other woman!"

"Yes—from a real bad one."

Sir Claude at least, the child could see, was not mystified; so little indeed that a smile of intelligence broke afresh in his eyes. He turned them in vague discomfort to Maisie, and then something in the way she met them caused him to chuck her playfully under the chin. It was not till after this that he good-naturedly met Mrs. Wix. "You think me much worse than I am."

"If that were true," she returned, "I would n't appeal to you. I do, Sir Claude, in the name of all that's good in you—and oh so earnestly! We can help each other. What you'll do for our young friend here I need n't say. That is n't even what I want to speak of now. What I want to speak of is what you'll *get*—don't you see?—from such an opportunity to take hold. Take hold of *us*—take hold of *her*. Make her your duty—make her your life: she'll repay you a thousand-fold!"

It was to Mrs. Wix, during this appeal, that Maisie's contemplation transferred itself: partly because, though her heart was in her throat for trepidation, her delicacy deterred her from appearing herself to press the question; partly from the coercion of seeing Mrs. Wix come out as Mrs. Wix had never come before—not even on the day of her call at

Mrs. Beale's with the news of mamma's marriage. On that day Mrs. Beale had surpassed her in dignity, but nobody could have surpassed her now. There was in fact at this moment a fascination for her pupil in the hint she seemed to give that she had still more of that surprise behind. So the sharpened sense of spectatorship was the child's main support, the long habit, from the first, of seeing herself in discussion and finding in the fury of it—she had had a glimpse of the game of football—a sort of compensation for the doom of a peculiar passivity. It gave her often an odd air of being present at her history in as separate a manner as if she could only get at experience by flattening her nose against a pane of glass. Such she felt to be the application of her nose while she waited for the effect of Mrs. Wix's eloquence. Sir Claude, however, did n't keep her long in a position so ungraceful: he sat down and opened his arms to her as he had done the day he came for her at her father's, and while he held her there, looking at her kindly, but as if their companion had brought the blood a good deal to his face, he said:

"Dear Mrs. Wix is magnificent, but she 's rather too grand about it. I mean the situation is n't after all quite so desperate or quite so simple. But I give you my word before her, and I give it to her before you, that I 'll never, never, forsake you. Do you hear that, old fellow, and do you take it in? I 'll stick to you through everything."

Maisie did take it in—took it with a long tremor of all her little being; and then as, to emphasise it, he drew her closer she buried her head on his shoulder and cried without sound and without pain. While she was so engaged she became aware that his own breast was agitated, and gathered from it with rapture that his tears were as silently flowing. Presently she heard a loud sob from Mrs. Wix—Mrs. Wix was the only one who made a noise.

She was to have made, for some time, none other but this, though within a few days, in conversation with her pupil, she described her intercourse with Ida as little better than the state of being battered. There was as yet nevertheless no attempt to eject her by force, and she recognised that Sir

Claude, taking such a stand as never before, had intervened with passion and with success. As Maisie remembered— and remembered wholly without disdain—that he had told her he was afraid of her ladyship, the little girl took this act of resolution as a proof of what, in the spirit of the engagement sealed by all their tears, he was really prepared to do. Mrs. Wix spoke to her of the pecuniary sacrifice by which she herself purchased the scant security she enjoyed and which, if it was a defence against the hand of violence, yet left her exposed to incredible rudeness. Did n't her ladyship find every hour of the day some artful means to humiliate and trample upon her? There was a quarter's salary owing her— a great name, even Maisie could suspect, for a small matter; she should never see it as long as she lived, but keeping quiet about it put her ladyship, thank heaven, a little in one's power. Now that he was doing so much else she could never have the grossness to apply for it to Sir Claude. He had sent home for schoolroom consumption a huge frosted cake, a wonderful delectable mountain with geological strata of jam, which might, with economy, see them through many days of their siege; but it was none the less known to Mrs. Wix that his affairs were more and more involved, and her fellow partaker looked back tenderly, in the light of these involutions, at the expression of face with which he had greeted the proposal that he should set up another establishment. Maisie felt that if their maintenance should hang by a thread they must still demean themselves with the highest delicacy. What he was doing was simply acting without delay, so far as his embarrassments permitted, on the inspiration of his elder friend. There was at this season a wonderful month of May—as soft as a drop of the wind in a gale that had kept one awake—when he took out his stepdaughter with a fresh alacrity and they rambled the great town in search, as Mrs. Wix called it, of combined amusement and instruction.

They rode on the top of 'buses; they visited out-lying parks; they went to cricket-matches where Maisie fell asleep; they tried a hundred places for the best one to have tea. This was his direct way of rising to Mrs. Wix's grand lesson—of making his little accepted charge his duty and his life. They dropped, under incontrollable impulses, into shops that they agreed were too big, to look at things that they agreed were too small, and it was during these hours that Mrs. Wix, alone at home, but a subject of regretful reference as they pulled off their gloves for refreshment, subsequently described herself as least sheltered from the blows her ladyship had achieved such ingenuity in dealing. She again and again repeated that she would n't so much have minded having her "attainments" held up to scorn and her knowledge of every subject denied, had n't she been branded as "low" in character and tone. There was by this time no pretence on the part of any one of denying it to be fortunate that her ladyship habitually left London every Saturday and was more and more disposed to a return late in the week. It was almost equally public that she regarded as a preposterous "pose," and indeed as a direct insult to herself, her husband's attitude of staying behind to look after a child for whom the most elaborate provision had been made. If there was a type Ida despised, Sir Claude communicated to Maisie, it was the man who pottered about town of a Sunday; and he also mentioned how often she had declared to him that if he had a grain of spirit he would be ashamed to accept a menial position about Mr. Farange's daughter. It was her ladyship's contention that he was in craven fear of his predecessor—otherwise he would recognise it as an obligation of plain decency to protect his wife against the outrage of that person's barefaced attempt to swindle her. The swindle was that Mr. Farange put upon her the whole intolerable burden; "and even when I pay for you myself," Sir Claude averred to his young friend, "she accuses me the more of truckling and grovelling." It was Mrs. Wix's conviction, they both knew, arrived at on independent grounds, that Ida's weekly excursions were feelers for a more considerable absence. If she came back later each week the week would be sure to arrive when she would n't come back at all. This appearance had of course much to do with Mrs. Wix's actual valour. Could they but hold out long enough the snug little home with Sir Claude would find itself informally established.

XIII

THIS might moreover have been taken to be the sense of a remark made by her stepfather as—one rainy day when the streets were all splash and two umbrellas unsociable and the wanderers had sought shelter in the National Gallery— Maisie sat beside him staring rather sightlessly at a roomful of pictures which he had mystified her much by speaking of with a bored sigh as a "silly superstition." They represented, with patches of gold and cataracts of purple, with stiff saints and angular angels, with ugly Madonnas and uglier babies, strange prayers and prostrations; so that she at first took his words for a protest against devotional idolatry—all the more that he had of late often come with her and with Mrs. Wix to morning church, a place of worship of Mrs. Wix's own choosing, where there was nothing of that sort; no haloes on heads, but only, during long sermons, beguiling backs of bonnets, and where, as her governess always afterwards observed, he gave the most earnest attention. It presently appeared, however, that his reference was merely to the affectation of admiring such ridiculous works—an admonition that she received from him as submissively as she received everything. What turn it gave to their talk need n't here be recorded: the transition to the colourless school-room and lonely Mrs. Wix was doubtless an effect of relaxed interest in what was before them. Maisie expressed in her own way the truth that she never went home nowadays without expecting to find the temple of her studies empty and the poor priestess cast out. This conveyed a full appreciation of her peril, and it was in rejoinder that Sir Claude uttered, acknowledging the source of that peril, the reassurance at which I have glanced. "Don't be afraid, my dear: I've squared her." It required indeed a supplement when he saw that it left the child momentarily blank. "I mean that your mother lets me do what I want so long as I let her do what *she* wants."

"So you *are* doing what you want?" Maisie asked.

"Rather, Miss Farange!"

Miss Farange turned it over. "And she's doing the same?"

"Up to the hilt!"

Again she considered. "Then, please, what may it be?"

"I would n't tell you for the whole world."

She gazed at a gaunt Madonna; after which she broke into a slow smile. "Well, I don't care, so long as you do let her."

"Oh you monster!"—and Sir Claude's gay vehemence brought him to his feet.

Another day, in another place—a place in Baker Street where at a hungry hour she had sat down with him to tea and buns—he brought out a question disconnected from previous talk. "I say, you know, what do you suppose your father *would* do?"

Maisie had n't long to cast about or to question his pleasant eyes. "If you were really to go with us? He'd make a great complaint."

He seemed amused at the term she employed. "Oh I should n't mind a 'complaint'!"

"He'd talk to every one about it," said Maisie.

"Well, I should n't mind that either."

"Of course not," the child hastened to respond. "You've told me you're not afraid of him."

"The question is are you?" said Sir Claude.

Maisie candidly considered; then she spoke resolutely. "No, not of papa."

"But of somebody else?"

"Certainly, of lots of people."

"Of your mother first and foremost of course."

"Dear, yes; more of mamma than of—than of—"

"Than of what?" Sir Claude asked as she hesitated for a comparison.

She thought over all objects of dread. "Than of a wild elephant!" she at last declared. "And you are too," she reminded him as he laughed.

"Oh yes, I am too."

Again she meditated. "Why then did you marry her?"

"Just because I *was* afraid."

"Even when she loved you?"

"That made her the more alarming."

For Maisie herself, though her companion seemed to find it droll, this opened up depths of gravity. "More alarming than she is now?"

"Well, in a different way. Fear, unfortunately, is a very big thing, and there's a great variety of kinds."

She took this in with complete intelligence. "Then I think I've got them all."

"You?" her friend cried. "Nonsense! You're thoroughly 'game.'"

"I'm awfully afraid of Mrs. Beale," Maisie objected.

He raised his smooth brows. "That charming woman?"

"Well," she answered, "you can't understand it because you're not in the same state."

She had been going on with a luminous "But" when, across the table, he laid his hand on her arm. "I *can* understand it," he confessed. "I *am* in the same state."

"Oh but she likes you so!" Maisie promptly pleaded.

Sir Claude literally coloured. "That has something to do with it."

Maisie wondered again. "Being liked with being afraid?"

"Yes, when it amounts to adoration."

"Then why aren't you afraid of *me*?"

"Because with you it amounts to that?" He had kept his hand on her arm. "Well, what prevents is simply that you're the gentlest spirit on earth. Besides—" he pursued; but he came to a pause.

"Besides—?"

"I *should* be in fear if you were older—there! See—you already make me talk nonsense," the young man added.

"The question's about your father. Is he likewise afraid of Mrs. Beale?"

"I think not. And yet he loves her," Maisie mused.

"Oh no—he does n't; not a bit!" After which, as his companion stared, Sir Claude apparently felt that he must make this oddity fit with her recollections. "There's nothing of that sort *now*."

But Maisie only stared the more. "They've changed?"

"Like your mother and me."

She wondered how he knew. "Then you've seen Mrs. Beale again?"

He demurred. "Oh no. She has written to me," he presently subjoined. "*She's* not afraid of your father either. No one at all is—really." Then he went on while Maisie's little mind, with its filial spring too relaxed from of old for a pang at this want of parental majesty, speculated on the vague relation between Mrs. Beale's courage and the question, for Mrs. Wix and herself, of a neat lodging with their friend. "She would n't care a bit if Mr. Farange should make a row."

"Do you mean about you and me and Mrs. Wix? Why should she care? It would n't hurt *her*."

Sir Claude, with his legs out and his hand diving into his trousers-pocket, threw back his head with a laugh just perceptibly tempered, as she thought, by a sigh. "My dear step-child, you're delightful! Look here, we must pay. You've had five buns?"

"How *can* you?" Maisie demanded, crimson under the eye of the young woman who had stepped to their board. "I've had three."

Shortly after this Mrs. Wix looked so ill that it was to be feared her ladyship had treated her to some unexampled passage. Maisie asked if anything worse than usual had occurred; whereupon the poor woman brought out with infinite gloom: "He has been seeing Mrs. Beale."

"Sir Claude?" The child remembered what he had said. "Oh no—not *seeing* her!"

"I beg your pardon. I absolutely know it." Mrs. Wix was as positive as she was dismal.

Maisie nevertheless ventured to challenge her. "And how, please, do you know it?"

She faltered a moment. "From herself. I've been to see her." Then on Maisie's visible surprise: "I went yesterday while you were out with him. He has seen her repeatedly."

It was not wholly clear to Maisie why Mrs. Wix should be prostrate at this discovery; but her general consciousness of the way things could be both perpetrated and resented always eased off for her the strain of the particular mystery. "There may be some mistake. He says he has n't."

Mrs. Wix turned paler, as if this were a still deeper ground for alarm. "He says so?—he denies that he has seen her?"

"He told me so three days ago. Perhaps she's mistaken," Maisie suggested.

"Do you mean perhaps she lies? She lies whenever it suits her, I'm very sure. But I know when people lie—and that's what I've loved in you, that *you* never do. Mrs. Beale did n't yesterday at any rate. He *has* seen her."

Maisie was silent a little. "He says not," she then repeated. "Perhaps—perhaps—" Once more she paused.

"Do you mean perhaps *he* lies?"

"Gracious goodness, no!" Maisie shouted.

Mrs. Wix's bitterness, however, again overflowed. "He does, he does," she cried, "and it's that that's just the worst of it! They'll take you, they'll take you, and what in the world will then become of me?" She threw herself afresh upon her pupil and wept over her with the inevitable effect of causing the child's own tears to flow. But Maisie could n't have told you if she had been crying at the image of their separation or at that of Sir Claude's untruth. As regards this deviation it was agreed between them that they were not in a position to bring it home to him. Mrs. Wix was in dread of doing anything to make him, as she said, "worse"; and Maisie was sufficiently initiated to be able to reflect that in speaking to her as he had done he had only wished to be tender of Mrs. Beale. It fell in with all her inclinations to think of him as tender, and she forbore to let him know that the two ladies had, as *she* would never do, betrayed him.

She had not long to keep her secret, for the next day, when she went out with him, he suddenly said in reference to some errand he had first proposed: "No, we won't do that—we'll do something else." On this, a few steps from the door, he stopped a hansom and helped her in; then following her he gave the driver over the top an address that she lost. When he was seated beside her she asked him where they were going; to which he replied "My dear child, you'll see." She saw while she watched and wondered that they took the direction of the Regent's Park; but she did n't know why he should make a mystery of that, and it was not till they passed under a pretty arch and drew up at a white house in a terrace from which the view, she thought, must be lovely that, mystified, she clutched him and broke out: "I shall see papa?"

He looked down at her with a kind smile. "No, probably not. I have n't brought you for that."

"Then whose house is it?"

"It's your father's. They've moved here."

She looked about: she had known Mr. Farange in four or five houses, and there was nothing astonishing in this except that it was the nicest place yet. "But I shall see Mrs. Beale?"

"It's to see her that I brought you."

She stared, very white, and, with her hand on his arm, though they had stopped, kept him sitting in the cab. "To leave me, do you mean?"

He could scarce bring it out. "It's not for me to say if you *can* stay. We must look into it."

"But if I do I shall see papa?"

"Oh some time or other, no doubt." Then Sir Claude went on: "Have you really so very great a dread of that?"

Maisie glanced away over the apron of the cab—gazed a minute at the green expanse of the Regent's Park and, at this moment colouring to the roots of her hair, felt the full, hot rush of an emotion more mature than any she had yet known. It consisted of an odd unexpected shame at placing in an inferior light, to so perfect a gentleman and so charming a person as Sir Claude, so very near a relative as Mr.

Farange. She remembered, however, her friend's telling her that no one was seriously afraid of her father, and she turned round with a small toss of her head. "Oh I dare say I can manage him!"

Sir Claude smiled, but she noted that the violence with which she had just changed colour had brought into his own face a slight compunctious and embarrassed flush. It was as if he had caught his first glimpse of her sense of responsibility. Neither of them made a movement to get out, and after an instant he said to her: "Look here, if you say so we won't after all go in."

"Ah but I want to see Mrs. Beale!" the child gently wailed.

"But what if she does decide to take you? Then, you know, you'll have to remain."

Maisie turned it over. "Straight on—and give you up?"

"Well—I don't quite know about giving me up."

"I mean as I gave up Mrs. Beale when I last went to mamma's. I could n't do without you here for anything like so long a time as that." It struck her as a hundred years since she had seen Mrs. Beale, who was on the other side of the door they were so near and whom she yet had not taken the jump to clasp in her arms.

"Oh I dare say you'll see more of me than you've seen of Mrs. Beale. It is n't in *me* to be so beautifully discreet," Sir Claude said. "But all the same," he continued, "I leave the thing, now that we're here, absolutely *with* you. You must settle it. We'll only go in if you say so. If you don't say so we'll turn right round and drive away."

"So in that case Mrs. Beale won't take me?"

"Well—not by any act of ours."

"And I shall be able to go on with mamma?" Maisie asked.

"Oh I don't say that!"

She considered. "But I thought you said you had squared her?"

Sir Claude poked his stick at the splashboard of the cab. "Not, my dear child, to the point she now requires."

"Then if she turns me out and I don't come here—?"

Sir Claude promptly took her up. "What do I offer you, you naturally enquire? My poor chick, that's just what I ask myself. I don't see it, I confess, quite as straight as Mrs. Wix."

His companion gazed a moment at what Mrs. Wix saw. "You mean *we* can't make a little family?"

"It's very base of me, no doubt, but I can't wholly chuck your mother."

Maisie, at this, emitted a low but lengthened sigh, a slight sound of reluctant assent which would certainly have been amusing to an auditor. "Then there is n't anything else?"

"I vow I don't quite see what there is."

Maisie waited; her silence seemed to signify that she too had no alternative to suggest. But she made another appeal. "If I come here you'll come to see me?"

"I won't lose sight of you."

"But how often will you come?" As he hung fire she pressed him. "Often and often?"

Still he faltered. "My dear old woman—" he began. Then he paused again, going on the next moment with a change of tone. "You're too funny! Yes then," he said; "often and often."

"All right!" Maisie jumped out. Mrs. Beale was at home, but not in the drawing-room, and when the butler had gone for her the child suddenly broke out: "But when I'm here what will Mrs. Wix do?"

"Ah you should have thought of that sooner!" said her companion with the first faint note of asperity she had ever heard him sound.

XIV

MRS. BEALE fairly swooped upon her, and the effect of the whole hour was to show the child how much, how quite formidably indeed, after all, she was loved. This was the more the case as her stepmother, so changed—in the very manner of her mother—that she really struck her as a new acquaintance, somehow recalled more familiarity than Maisie could feel. A rich strong expressive affection in short pounced upon her in the shape of a handsomer, ampler, older Mrs. Beale. It was like making a fine friend, and they had n't been a minute together before she felt elated at the way she had met the choice imposed on her in the cab. There was a whole future in the combination of Mrs. Beale's beauty and Mrs. Beale's hug. She seemed to Maisie charming to behold, and also to have no connexion at all with anybody who had once mended underclothing and had meals in the nursery. The child knew one of her father's wives was a woman of fashion, but she had always dimly made a distinction, not applying that epithet without reserve to the other. Mrs. Beale had since their separation acquired a conspicuous right to it, and Maisie's first flush of response to her present delight coloured all her splendour with meanings that this time were sweet. She had told Sir Claude she was afraid of the lady in the Regent's Park; but she had confidence enough to break, on the spot, into the frankest appreciation. "Why, are n't you beautiful? Is n't she beautiful, Sir Claude, *is n't* she?"

"The handsomest woman in London, simply," Sir Claude gallantly replied. "Just as sure as you 're the best little girl!"

Well, the handsomest woman in London gave herself up, with tender lustrous looks and every demonstration of fondness, to a happiness at last clutched again. There was almost as vivid a bloom in her maturity as in mamma's, and it took her but a short time to give her little friend an impression of positive power—an impression that seemed to begin like a long bright day. This was a perception on Maisie's part that neither mamma, nor Sir Claude, nor Mrs. Wix, with their immense and so varied respective attractions, had exactly kindled, and that made an immediate difference when the talk, as it promptly did, began to turn to her father. Oh yes, Mr. Farange was a complication, but she saw now that he would n't be one for his daughter. For Mrs. Beale certainly he was an immense one—she speedily made known as much;

but Mrs. Beale from this moment presented herself to Maisie as a person to whom a great gift had come. The great gift was just for handling complications. Maisie felt how little she made of them when, after she had dropped to Sir Claude some recall of a previous meeting, he made answer, with a sound of consternation and yet an air of relief, that he had denied to their companion their having, since the day he came for her, seen each other till that moment.

Mrs. Beale could but vaguely pity it. "Why did you do anything so silly?"

"To protect your reputation."

"From Maisie?" Mrs. Beale was much amused. "My reputation with Maisie is too good to suffer."

"But you believed me, you rascal, did n't you?" Sir Claude asked of the child.

She looked at him; she smiled. "Her reputation did suffer. I discovered you had been here."

He was not too chagrined to laugh. "The way, my dear, you talk of that sort of thing!"

"How should she talk," Mrs. Beale wanted to know, "after all this wretched time with her mother?"

"It was not mamma who told me," Maisie explained. "It was only Mrs. Wix." She was hesitating whether to bring out before Sir Claude the source of Mrs. Wix's information; but Mrs. Beale, addressing the young man, showed the vanity of scruples.

"Do you know that preposterous person came to see me a day or two ago?—when I told her I had seen you repeatedly."

Sir Claude, for once in a way, was disconcerted. "The old cat! She never told me. Then you thought I had lied?" he demanded of Maisie.

She was flurried by the term with which he had qualified her gentle friend, but she took the occasion for one to which she must in every manner lend herself. "Oh I did n't mind! But Mrs. Wix did," she added with an intention benevolent to her governess.

Her intention was not very effective as regards Mrs. Beale. "Mrs. Wix is too idiotic!" that lady declared.

"But to you, of all people," Sir Claude asked, "what had she to say?"

"Why that, like Mrs. Micawber—whom she must, I think, rather resemble—she will never, never, never desert Miss Farange."

"Oh I'll make that all right!" Sir Claude cheerfully returned.

"I'm sure I hope so, my dear man," said Mrs. Beale, while Maisie wondered just how he would proceed. Before she had time to ask Mrs. Beale continued: "That's not all she came to do, if you please. But you'll never guess the rest."

"Shall *I* guess it?" Maisie quavered.

Mrs. Beale was again amused. "Why you're just the person! It must be quite the sort of thing you've heard at your awful mother's. Have you never seen women there crying to her to 'spare' the men they love?"

Maisie, wondering, tried to remember; but Sir Claude was freshly diverted. "Oh they don't trouble about Ida! Mrs. Wix cried to you to spare *me*?"

"She regularly went down on her knees to me."

"The darling old dear!" the young man exclaimed.

These words were a joy to Maisie—they made up for his previous description of Mrs. Wix. "And *will* you spare him?" she asked of Mrs. Beale.

Her stepmother, seizing her and kissing her again, seemed charmed with the tone of her question. "Not an inch of him! I'll pick him to the bone!"

"You mean that he'll really come often?" Maisie pressed.

Mrs. Beale turned lovely eyes to Sir Claude. "That's not for me to say—it's for him."

He said nothing at once, however; with his hands in his pockets and vaguely humming a tune—even Maisie could see he was a little nervous—he only walked to the window and looked out at the Regent's Park. "Well, he has promised, Maisie said. "But how will papa like it?"

"His being in and out? Ah that's a question that, to be frank with you, my dear, hardly matters. In point of fact, however, Beale greatly enjoys the idea that Sir Claude too, poor man, has been forced to quarrel with your mother."

Sir Claude turned round and spoke gravely and kindly. "Don't be afraid, Maisie; you won't lose sight of me."

"Thank you so much!" Maisie was radiant. "But what I meant—don't you know?—was what papa would say to *me*."

"Oh I've been having that out with him," said Mrs. Beale. "He'll behave well enough. You see the great difficulty is that, though he changes every three days about everything else in the world, he has never changed about your mother. It's a caution, the way he hates her."

Sir Claude gave a short laugh. "It certainly can't beat the way she still hates *him*!"

"Well," Mrs. Beale went on obligingly, "nothing can take the place of that feeling with either of them, and the best way they can think of to show it is for each to leave you as long as possible on the hands of the other. There's nothing, as you've seen for yourself, that makes either so furious. It is n't, asking so little as you do, that you're much of an expense or a trouble; it's only that you make each feel so well how nasty the other wants to be. Therefore Beale goes on loathing your mother too much to have any great fury left for any one else. Besides, you know, I've squared him."

"Oh Lord!" Sir Claude cried with a louder laugh and turning again to the window.

"*I* know how!" Maisie was prompt to proclaim. "By letting him do what he wants on condition that he lets you also do it."

"You're too delicious, my own pet!"—she was involved in another hug. "How in the world have I got on so long without you? I've not been happy, love," said Mrs. Beale with her cheek to the child's.

"Be happy now!"—Maisie throbbed with shy tenderness.

"I think I shall be. You'll save me."

"As I'm saving Sir Claude?" the little girl asked eagerly.

Mrs. Beale, a trifle at a loss, appealed to her visitor, "Is she really?"

He showed high amusement at Maisie's question. "It's dear Mrs. Wix's idea. There may be something in it."

"He makes me his duty—he makes me his life," Maisie set forth to her stepmother.

"Why that's what *I* want to do!"—Mrs. Beale, so anticipated, turned pink with astonishment.

"Well, you can do it together. Then he'll *have* to come!"

Mrs. Beale by this time had her young friend fairly in her lap and she smiled up at Sir Claude. "Shall we do it together?"

His laughter had dropped, and for a moment he turned his handsome serious face not to his hostess, but to his stepdaughter. "Well, it's rather more decent than some things. Upon my soul, the way things are going, it seems to me the only decency!" He had the air of arguing it out to Maisie, of presenting it, through an impulse of conscience, as a connexion in which they could honourably see her participate; though his plea of mere "decency" might well have appeared to fall below her rosy little vision. "If we're not good for *you*," he exclaimed, "I'll be hanged if I know who we shall be good for!"

Mrs. Beale showed the child an intenser light. "I dare say you *will* save us—from one thing and another."

"Oh I know what she'll save *me* from!" Sir Claude roundly asserted. "There'll be rows of course," he went on.

Mrs. Beale quickly took him up. "Yes, but they'll be nothing—for you at least—to the rows your wife makes as it is. I can bear what *I* suffer—I can't bear what you go through."

"We're doing a good deal for you, you know, young woman," Sir Claude went on to Maisie with the same gravity.

She coloured with a sense of obligation and the eagerness of her desire it should be remarked how little was lost on her. "Oh I know!"

"Then you must keep us all right!" This time he laughed.

"How you talk to her!" cried Mrs. Beale.

"No worse than you!" he gaily answered.

"Handsome is that handsome does!" she returned in the same spirit. "You can take off your things," she went on, releasing Maisie.

The child, on her feet, was all emotion. "Then I'm just to stop—this way?"

"It will do as well as any other. Sir Claude, tomorrow, will have your things brought."

"I'll bring them myself. Upon my word I'll see them packed!" Sir Claude promised. "Come here and unbutton."

He had beckoned his young companion to where he sat, and he helped to disengage her from her coverings while Mrs. Beale, from a little distance, smiled at the hand he displayed. "There's a stepfather for you! I'm bound to say, you know, that he makes up for the want of other people."

"He makes up for the want of a nurse!" Sir Claude laughed. "Don't you remember I told you so the very first time?"

"Remember? It was exactly what made me think so well of you!"

"Nothing would induce me," the young man said to Maisie, "to tell you what made me think so well of *her*." Having divested the child he kissed her gently and gave her a little pat to make her stand off. The pat was accompanied with a vague sigh in which his gravity of a moment before came back. "All the same, if you had n't had the fatal gift of beauty—!"

"Well, what?" Maisie asked, wondering why he paused. It was the first time she had heard of her beauty.

"Why, we should n't all be thinking so well of each other!"

"He is n't speaking of personal loveliness—you've not *that* vulgar beauty, my dear, at all," Mrs. Beale explained. "He's just talking of plain dull charm of character."

"Her character's the most extraordinary thing in all the world," Sir Claude stated to Mrs. Beale.

"Oh I know all about that sort of thing!"—she fairly bridled with the knowledge.

It gave Maisie somehow a sudden sense of responsibility from which she sought refuge. "Well, you've got it too, 'that sort of thing'—you've got the fatal gift: you both really have!" she broke out.

"Beauty of character? My dear boy, we have n't a pennyworth!" Sir Claude protested.

"Speak for yourself, sir!" leaped lightly from Mrs. Beale. "I'm good and I'm clever. What more do you want? For you, I'll spare your blushes and not be personal— I'll simply say that you're as handsome as you can stick together."

"You're both very lovely; you can't get out of it!"— Maisie felt the need of carrying her point. "And it's beautiful to see you side by side."

Sir Claude had taken his hat and stick; he stood looking at her a moment. "You're a comfort in trouble! But I must go home and pack you."

"And when will you come back?—to-morrow, to-morrow?"

"You see what we're in for!" he said to Mrs. Beale.

"Well, I can bear it if you can."

Their companion gazed from one of them to the other, thinking that though she had been happy indeed between Sir Claude and Mrs. Wix she should evidently be happier still between Sir Claude and Mrs. Beale. But it was like being perched on a prancing horse, and she made a movement to hold on to something. "Then, you know, shan't I bid good-bye to Mrs. Wix?"

"Oh I'll make it all right with her," said Sir Claude.

Maisie considered. "And with mamma?"

"Ah mamma!" he sadly laughed.

Even for the child this was scarcely ambiguous; but Mrs. Beale •endeavoured to contribute to its clearness. "Your mother will crow, she'll crow—"

"Like the early bird!" said Sir Claude as she looked about for a comparison.

"She'll need no consolation," Mrs. Beale went on, "for having made your father grandly blaspheme."

Maisie stared. "Will he grandly blaspheme?" It was impressive, it might have been out of the Bible, and her question produced a fresh play of caresses, in which Sir Claude also engaged. She wondered meanwhile who, if Mrs. Wix was disposed of, would represent in her life the element of geography and anecdote; and she presently surmounted the delicacy she felt about asking. "Won't there be any one to give me lessons?"

Mrs. Beale was prepared with a reply that struck her as absolutely magnificent. "You shall have such lessons as you've never had in all your life. You shall go to courses."

"Courses?" Maisie had never heard of such things.

"At institutions—on subjects."

Maisie continued to stare. "Subjects?"

Mrs. Beale was really splendid. "All the most important ones. French literature—and sacred history. You'll take part in classes—with awfully smart children."

"I'm going to look thoroughly into the whole thing, you know." And Sir Claude, with characteristic kindness, gave her a nod of assurance accompanied by a friendly wink.

But Mrs. Beale went much further. "My dear child, you shall attend lectures."

The horizon was suddenly vast and Maisie felt herself the smaller for it. "All alone?"

"Oh no; I'll attend them with you," said Sir Claude. "They'll teach me a lot I don't know."

"So they will me," Mrs. Beale gravely admitted. "We'll go with her together—it will be charming. It's ages," she confessed to Maisie, "since I've had any time for study. That's another sweet way in which you'll be a motive to us. Oh won't the good she'll do us be immense?" she broke out uncontrollably to Sir Claude.

He weighed it; then he replied: "That's certainly our idea." Of this idea Maisie naturally had less of a grasp, but it inspired her with almost equal enthusiasm. If in so bright a prospect there would be nothing to long for it followed that she wouldn't long for Mrs. Wix; but her consciousness of her assent to the absence of that fond figure caused a pair of words that had often sounded in her ears to ring in them again. It showed her in short what her father had always meant by calling her mother a "low sneak" and her mother by calling her father one. She wondered if she herself should n't be a low sneak in learning to be so happy without Mrs. Wix. What would Mrs. Wix do?—where would Mrs. Wix go? Before Maisie knew it, and at the door, as Sir Claude was off, these anxieties, on her lips, grew articulate and her stepfather had stopped long enough to answer them. "Oh I'll square her!" he cried; and with this he departed.

Face to face with Mrs. Beale Maisie, giving a sigh of relief, looked round at what seemed to her the dawn of a higher order. "Then *every one* will be squared!" she peacefully said. On which her stepmother affectionately bent over her again.

XV

IT was Susan Ash who came to her with the news: "He's downstairs, miss, and he do look beautiful."

In the schoolroom at her father's, which had pretty blue curtains, she had been making out at the piano a lovely little thing, as Mrs. Beale called it, a "Moon-light Berceuse" sent her through the post by Sir Claude, who considered that her musical education had been deplorably neglected and who, the last months at her mother's, had been on the point of making arrangements for regular lessons. She knew from him familiarly that the real thing, as he said, was shockingly dear and that anything else was a waste of money, and she therefore rejoiced the more at the sacrifice represented by this composition, of which the price, five shillings, was marked on the cover and which was evidently the real thing. She was already on her feet. "Mrs. Beale has sent up for me?"

"Oh no—it's not that," said Susan Ash. "Mrs. Beale has been out this hour."

"Then papa!"

"Dear no—not papa. You'll do, miss, all but them wandering 'airs," Susan went on. "Your papa never came 'ome at all," she added.

"Home from where?" Maisie responded a little absently and very excitedly. She gave a wild manual brush to her locks.

"Oh that, miss, I should be very sorry to tell you! I'd rather tuck away that white thing behind—though I'm blest if it's my work."

"Do then, please. I know where papa was," Maisie impatiently continued.

"Well, in your place I would n't tell."

"He was at the club—the Chrysanthemum. So!"

"All night long? Why the flowers shut up at night, you know!" cried Susan Ash.

"Well, I don't care"—the child was at the door. "Sir Claude asked for me *alone*?"

"The same as if you was a duchess."

Maisie was aware on her way downstairs that she was now quite as happy as one, and also, a moment later, as she hung round his neck, that even such a personage would scarce commit herself more grandly. There was moreover a hint of the duchess in the infinite point with which, as she felt, she exclaimed: "And this is what you call coming *often*?"

Sir Claude met her delightfully and in the same fine spirit. "My dear old man, don't make me a scene—I assure you it's what every woman I look at does. Let us have some fun—it's a lovely day: clap on something smart and come out with me; then we'll talk it over quietly." They were on their way five minutes later to Hyde Park, and nothing that even in the good days at her mother's they had ever talked over had more of the sweetness of tranquillity than his present prompt explanations. He was at his best in such an office and with the exception of Mrs. Wix the only person she had met in her life who ever explained. With him, however, the act had an authority transcending the wisdom of woman. It all came back—all the plans that always failed, all the rewards and bribes that she was perpetually paying for in advance and perpetually out of pocket by afterwards— the whole great stress to be dealt with introduced her on each occasion afresh to the question of money. Even she herself almost knew how it would have expressed the strength of his empire to say that to shuffle away her sense of being duped he had only, from under his lovely moustache, to breathe upon it. It was somehow in the nature of plans to be expensive and in the nature of the expensive to be impossible. To be "involved" was of the essence of everybody's affairs, and also at every particular moment to be more involved than usual. This had been the case with Sir Claude's, with papa's, with mamma's, with Mrs. Beale's and with Maisie's own at the particular moment, a moment of several weeks, that had elapsed since our young lady had been re-established at her father's. There was n't "two-and-tuppence" for anything or for any one, and that was why there had been no sequel to the classes in French literature with all the smart little girls. It was devilish awkward, did n't she see? to try, without even the limited capital mentioned, to mix her up with a remote array that glittered before her after this as the children of the rich. She was to feel henceforth as if she were flattening her nose upon the hard window-pane of the sweet-shop of knowledge. If the classes, however, that were select, and accordingly the only ones, were impossibly dear, the lectures at the institutions—at least at some of them—were directly addressed to the intelligent poor, and it therefore had to be easier still to produce on the spot the reason why she had been taken to none. This reason, Sir Claude said, was that she happened to be just going to be, though they had nothing to do with that in now directing their steps to the banks of the Serpentine. Maisie's own park, in the north, had been nearer at hand, but they rolled westward in a hansom because at the end of the sweet June days this was the direction taken by every one that any one looked at. They cultivated for an hour, on the Row and by the Drive, this opportunity for each observer to amuse and for one of them indeed, not a little hilariously, to mystify the other, and before the hour was over Maisie had elicited, in reply to her sharpest challenge, a further account of her friend's long absence.

"Why I've broken my word to you so dreadfully—

promising so solemnly and then never coming? Well, my dear, that's a question that, not seeing me day after day, you must very often have put to Mrs. Beale."

"Oh yes," the child replied; "again and again."

"And what has she told you?"

"That you're as bad as you're beautiful."

"Is that what she says?"

"Those very words."

"Ah the dear old soul!" Sir Claude was much diverted, and his loud, clear laugh was all his explanation. Those were just the words Maisie had last heard him use about Mrs. Wix. She clung to his hand, which was encased in a pearl-grey glove ornamented with the thick black lines that, at her mother's, always used to strike her as connected with the way the bestitched fists of the long ladies carried, with the elbows well out, their umbrellas upside down. The mere sense of his grasp in her own covered the ground of loss just as much as the ground of gain. His presence was like an object brought so close to her face that she could n't see round its edges. He himself, however, remained showman of the spectacle even after they had passed out of the Park and begun, under the charm of the spot and the season, to stroll in Kensington Gardens. What they had left behind them was, as he said, only a pretty bad circus, and, through prepossessing gates and over a bridge, they had come in a quarter of an hour, as he also remarked, a hundred miles from London. A great green glade was before them, and high old trees, and under the shade of these, in the fresh turf, the crooked course of a rural footpath. "It's the Forest of Arden," Sir Claude had just delightfully observed, "and I'm the banished duke, and you're—what was the young woman called?—the artless country wench. And there," he went on, "is the other girl—what's her name, Rosalind?—and (don't you know?) the fellow who was making up to her. Upon my word he *is* making up to her!"

His allusion was to a couple who, side by side, at the end of the glade, were moving in the same direction as themselves. These distant figures, in their slow stroll (which kept them so close together that their heads, drooping a little forward, almost touched), presented the back of a lady who looked tall, who was evidently a very fine woman, and that of a gentleman whose left hand appeared to be passed well into her arm while his right, behind him, made jerky motions with the stick that it grasped. Maisie's fancy responded for an instant to her friend's idea that the sight was idyllic; then, stopping short, she brought out with all her clearness: "Why mercy—if it is n't mamma!"

Sir Claude paused with a stare. "Mamma? But mamma's at Brussels."

Maisie, with her eyes on the lady, wondered. "At Brussels?"

"She's gone to play a match."

"At billiards? You did n't tell me."

"Of course I did n't!" Sir Claude ejaculated. "There's plenty I don't tell you. She went on Wednesday."

The couple had added to their distance, but Maisie's eyes more than kept pace with them. "Then she has come back."

Sir Claude watched the lady. "It's much more likely she never went!"

"It's mamma!" the child said with decision.

They had stood still, but Sir Claude had made the most of his opportunity, and it happened that just at this moment, at the end of the vista, the others halted and, still showing only their backs, seemed to stay talking. "Right you are, my duck!" he exclaimed at last. "It's my own sweet wife!"

He had spoken with a laugh, but he had changed colour, and Maisie quickly looked away from him. "Then who is it with her?"

"Blest if I know!" said Sir Claude.

"Is it Mr. Perriam?"

"Oh dear no—Perriam's smashed."

"Smashed?"

"Exposed—in the City. But there are quantities of others!" Sir Claude smiled.

Maisie appeared to count them; she studied the gentleman's back. "Then is this Lord Eric?"

For a moment her companion made no answer, and when

she turned her eyes again to him he was looking at her, she thought, rather queerly. "What do you know about Lord Eric?"

She tried innocently to be odd in return. "Oh I know more than you think! Is it Lord Eric?" she repeated.

"It may be. Blest if I care!"

Their friends had slightly separated and now, as Sir Claude spoke, suddenly faced round, showing all the splendour of her ladyship and all the mystery of her comrade. Maisie held her breath. "They're coming!"

"Let them come." And Sir Claude, pulling out his cigarettes, began to strike a light.

"We shall meet them!"

"No. They'll meet *us*."

Maisie stood her ground. "They see us. Just look."

Sir Claude threw away his match. "Come straight on." The others, in the return, evidently startled, had half-paused again, keeping well apart. "She's horribly surprised and wants to slope," he continued. "But it's too late."

Maisie advanced beside him, making out even across the interval that her ladyship was ill at ease. "Then what will she do?"

Sir Claude puffed his cigarette. "She's quickly thinking." He appeared to enjoy it.

Ida had wavered but an instant; her companion clearly gave her moral support. Maisie thought he somehow looked brave, and he had no likeness whatever to Mr. Perriam. His face, thin and rather sharp, was smooth, and it was not till they came nearer that she saw he had a remarkably fair little moustache. She could already see that his eyes were of the lightest blue. He was far nicer than Mr. Perriam. Mamma looked terrible from afar, but even under her guns the child's curiosity flickered and she appealed again to Sir Claude. "Is it—*is* it Lord Eric?"

Sir Claude smoked composedly enough. "I think it's the Count."

This was a happy solution—it fitted her idea of a count. But what idea, as she now came grandly on, did mamma fit?—unless that of an actress, in some tremendous situation, sweeping down to the footlights as if she would jump them. Maisie felt really so frightened that before she knew it she had passed her hand into Sir Claude's arm. Her pressure caused him to stop, and at the sight of this the other couple came equally to a stand and, beyond the diminished space, remained a moment more in talk. This, however, was the matter of an instant; leaving the Count apparently to come round more circuitously—an outflanking movement, if Maisie had but known—her ladyship resumed the onset. "What *will* she do now?" her daughter asked.

Sir Claude was at present in a position to say: "Try to pretend it's me."

"You?"

"Why that I'm up to something."

In another minute poor Ida had justified this prediction, erect there before them like a figure of justice in full dress. There were parts of her face that grew whiter while Maisie looked, and other parts in which this change seemed to make other colours reign with more intensity. "What are you doing with my daughter?" she demanded of her husband; in spite of the indignant tone of which Maisie had a greater sense than ever in her life before of not being personally noticed. It seemed to her Sir Claude also grew pale as an effect of the loud defiance with which Ida twice repeated this question. He put her, instead of answering it, an enquiry of his own: "Who the devil have you got hold of *now*?" and at this her ladyship turned tremendously to the child, glaring at her as at an equal plotter of sin. Maisie received in petrifaction the full force of her mother's huge painted eyes—they were like Japanese lanterns swung under festal arches. But life came back to her from a tone suddenly and strangely softened. "Go straight to that gentleman, my dear; I've asked him to take you a few minutes. He's charming—go. I've something to say to *this* creature."

Maisie felt Sir Claude immediately clutch her. "No, no—thank you: that won't do. She's mine."

"Yours?" It was confounding to Maisie to hear

her speak quite as if she had never heard of Sir Claude before.

"Mine. You've given her up. You've not another word to say about her. I have her from her father," said Sir Claude—a statement that startled his companion, who could also measure its lively action on her mother.

There was visibly, however, an influence that made Ida consider; she glanced at the gentleman she had left, who, having strolled with his hands in his pockets to some distance, stood there with unembarrassed vagueness. She directed to him the face that was like an illuminated garden, turnstile and all, for the frequentation of which he had his season-ticket; then she looked again at Sir Claude. "I've given her up to her father to *keep*—not to get rid of by sending about the town either with you or with any one else. If she's not to mind me let *him* come and tell me so. I decline to take it from another person, and I like your pretending that with your humbug of 'interest' you've a leg to stand on. I know your game and have something now to say to you about it."

Sir Claude gave a squeeze of the child's arm. "Did n't I tell you she'd have, Miss Farange?"

"You're uncommonly afraid to hear it," Ida went on; "but if you think she'll protect you from it you're mightily mistaken." She gave him a moment. "I'll give her the benefit as soon as look at you. Should you like her to know, my dear?" Maisie had a sense of her launching the question with effect; yet our young lady was also conscious of hoping that Sir Claude would declare that preference. We have already learned that she had come to like people's liking her to "know." Before he could reply at all, none the less, her mother opened a pair of arms of extraordinary elegance, and then she felt the loosening of his grasp. "My own child," Ida murmured in a voice—a voice of sudden confused tenderness—that it seemed to her she heard for the first time. She wavered but an instant, thrilled with the first direct appeal, as distinguished from the mere maternal pull, she had ever had from lips that, even in the old vociferous years, had always been sharp. The next moment she was on her mother's breast, where, amid a wilderness of trinkets, she felt as if she had suddenly been thrust, with a smash of glass, into a jeweller's shop-front, but only to be as suddenly ejected with a push and the brisk injunction: "Now go to the Captain!"

Maisie glanced at the gentleman submissively, but felt the want of more introduction. "The Captain?"

Sir Claude broke into a laugh. "I told her it was the Count."

Ida stared; she rose so superior that she was colossal. "You're too utterly loathsome," she then declared. "Be off!" she repeated to her daughter.

Maisie started, moved backward and, looking at Sir Claude, "Only for a moment," she signed to him in her bewilderment.

But he was too angry to heed her—too angry with his wife; as she turned away she heard his anger break out. "You damned old b——!" —she could n't quite hear all. It was enough, it was too much: she fled before it, rushing even to a stranger for the shock of such a change of tone.

XVI

As she met the Captain's light blue eyes the greatest marvel occurred; she felt a sudden relief at finding them reply with anxiety to the horror in her face. "What in the world has he done?" He put it all on Sir Claude.

"He has called her a damned old brute." She could n't help bringing that out.

The Captain, at the same elevation as her ladyship, gaped wide; then of course, like every one else, he was convulsed. But he instantly caught himself up, echoing her bad words. "A damned old brute—your mother?"

Maisie was already conscious of her second movement. "I think she tried to make him angry."

The Captain's stupefaction was fine. "Angry—*she*? Why she's an angel!"

On the spot, as he said this, his face won her over; it was so bright and kind, and his blue eyes had such a reflexion of some mysterious grace that, for him at least, her mother had put forth. Her fund of observation enabled her as she gazed up at him to place him: he was a candid simple soldier; very grave—she came back to that—but not at all terrible. At any rate he struck a note that was new to her and that after a moment made her say: "Do you like her very much?"

He smiled down at her, hesitating, looking pleasanter and pleasanter. "Let me tell you about your mother."

He put out a big military hand which she immediately took, and they turned off together to where a couple of chairs had been placed under one of the trees. "She told me to come to you," Maisie explained as they went; and presently she was close to him in a chair, with the prettiest of pictures—the sheen of the lake through other trees—before them, and the sound of birds, the plash of boats, the play of children in the air. The Captain, inclining his military person, sat sideways to be closer and kinder, and as her hand was on the arm of her seat he put his own down on it again to emphasise something he had to say that would be good for her to hear. He had already told her how her mother, from the moment of seeing her so unexpectedly with a person who was—well, not at all the right person, had promptly asked him to take charge of her while she herself tackled, as she said, the real culprit. He gave the child the sense of doing for the time what he liked with her; ten minutes before she had never seen him, but she could now sit there touching him, touched and impressed by him and thinking it nice when a gentleman was thin and brown— brown with a kind of clear depth that made his straw-coloured moustache almost white and his eyes resemble little pale flowers. The most extraordinary thing was the way she did n't appear just then to mind Sir Claude's being tackled. The Captain was n't a bit like him, for it was an odd part of the pleasantness of mamma's friend that it resided in a manner in this friend's having a face so informally put together that the only kindness could be to call it funny. An odder part still was that it finally made our young lady, to classify him further, say to herself that, of all people in the world, he reminded her most insidiously of Mrs. Wix. He had neither straighteners nor a diadem, nor, at least in the same place as the other, a button; he was sun-burnt and deep-voiced and smelt of cigars, yet he marvellously had more in common with her old governess than with her young step-father. What he had to say to her that was good for her to hear was that her poor mother (did n't she know?) was the best friend he had ever had in all his life. And he added: "She has told me ever so much about you. I'm awfully glad to know you."

She had never, she thought, been so addressed as a young lady, not even by Sir Claude the day, so long ago, that she found him with Mrs. Beale. It struck her as the way that at balls, by delightful partners, young ladies must be spoken to in the intervals of dances; and she tried to think of something that would meet it at the same high point. But this effort flurried her, and all she could produce was: "At first, you know, I thought you were Lord Eric."

The Captain looked vague. "Lord Eric?"

"And then Sir Claude thought you were the Count."

At this he laughed out. "Why he's only five foot high and as red as a lobster!" Maisie laughed, with a certain elegance, in return—the young lady at the ball certainly would—and was on the point, as conscientiously, of pursuing the subject with an agreeable question. But before she could speak her companion challenged her. "Who in the world's Lord Eric?"

"Don't you know him?" She judged her young lady would say that with light surprise.

"Do you mean a fat man with his mouth always open?" She had to confess that their acquaintance was so limited that she could only describe the bearer of the name as a friend of mamma's; but a light suddenly came to the

Captain, who quickly spoke as knowing her man. "What-do-you-call-him's brother, the fellow that owned Bobolink?" Then, with all his kindness, he contradicted her flat. "Oh dear no; your mother never knew *him*."

"But Mrs. Wix said so," the child risked.

"Mrs. Wix?"

"My old governess."

This again seemed amusing to the Captain. "She mixed him up, your old governess. He's an awful beast. Your mother never looked at him."

He was as positive as he was friendly, but he dropped for a minute after this into a silence that gave Maisie, confused but ingenious, a chance to redeem the mistake of pretending to know too much by the humility of inviting further correction. "And does n't she know the Count?"

"Oh I dare say! But he's another ass." After which abruptly, with a different look, he put down again on the back of her own the hand he had momentarily removed. Maisie even thought he coloured a little. "I want tremendously to speak to you. You must never believe any harm of your mother."

"Oh I assure you I *don't*!" cried the child, blushing, herself, up to her eyes in a sudden surge of deprecation of such a thought.

The Captain, bending his head, raised her hand to his lips with a benevolence that made her wish her glove had been nicer. "Of course you don't when you know how fond she is of *you*."

"She's fond of me?" Maisie panted.

"Tremendously. But she thinks you don't like her. You *must* like her. She has had too much to put up with."

"Oh yes—I know!" She rejoiced that she had never denied it.

"Of course I've no right to speak of her except as a particular friend," the Captain went on. "But she's a splendid woman. She has never had any sort of justice."

"Has n't she?"—his companion, to hear the words, felt a thrill altogether new.

"Perhaps I ought n't to say it to you, but she has had everything to suffer."

"Oh yes—you can *say* it to me!" Maisie hastened to profess.

The Captain was glad. "Well, you need n't tell. It's all for *you*—do you see?"

Serious and smiling she only wanted to take it from him. "It's between you and me! Oh there are lots of things I've never told!"

"Well, keep this with the rest. I assure you she has had the most infernal time, no matter what any one says to the contrary. She's the cleverest woman I ever saw in all my life. She's too charming." She had been touched already by his tone, and now she leaned back in her chair and felt something tremble within her. "She's tremendous fun—she can do all sorts of things better than I've ever seen any one. She has the pluck of fifty—and I know; I assure you I do. She has the nerve for a tiger-shoot—by Jove I'd *take* her! And she is awfully open and generous, don't you know? there are women that are such horrid sneaks. She'll go through anything for any one she likes." He appeared to watch for a moment the effect on his companion of this emphasis; then he gave a small sigh that mourned the limits of the speakable. But it was almost with the note of a fresh challenge that he wound up: "Look here, she's *true*!"

Maisie had so little desire to assert the contrary that she found herself, in the intensity of her response, throbbing with a joy still less utterable than the essence of the Captain's admiration. She was fairly hushed with the sense that he spoke of her mother as she had never heard any one speak. It came over her as she sat silent that, after all, this admiration and this respect were quite new words, which took a distinction from the fact that nothing in the least resembling them in quality had on any occasion dropped from the lips of her father, of Mrs. Beale, of Sir Claude or even of Mrs. Wix. What it appeared to her to come to was that on the subject of her ladyship it was the first real kindness she had heard, so that at the touch of it something strange and deep and pitying surged up within her—a revelation that, practically and so far as she knew, her mother, apart from this, had only been disliked. Mrs. Wix's original account of Sir Claude's affection seemed as empty now as the chorus in a children's game, and the husband and wife, but a little way off at that moment, were face to face in hatred and with the dreadful name he had called her still in the air. What was it the Captain on the other hand had called her? Maisie wanted to hear that again. The tears filled her eyes and rolled down her cheeks, which burned under them with the rush of a consciousness that for her too, five minutes before, the vivid towering beauty whose assault she awaited had been, a moment long, an object of pure dread. She became on the spot indifferent to her usual fear of showing what in children was notoriously most offensive—presented to her companion, soundlessly but hideously, her wet distorted face. She cried, with a pang, straight *at* him, cried as she had never cried at any one in all her life. "Oh do you love her?" she brought out with a gulp that was the effect of her trying not to make a noise.

It was doubtless another consequence of the thick mist through which she saw him that in reply to her question the Captain gave her such a queer blurred look. He stammered, yet in his voice there was also the ring of a great awkward insistence. "Of course I'm tremendously fond of her—I like her better than any woman I ever saw. I don't mind in the least telling you that," he went on, "and I should think myself a great beast if I did." Then to show that his position was superlatively clear he made her, with a kindness that even Sir Claude had never surpassed, tremble again as she had trembled at his first outbreak. He called her by her name, and her name drove it home. "My dear Maisie, your mother's an angel!"

It was an almost unbelievable balm—it soothed so her impression of danger and pain. She sank back in her chair, she covered her face with her hands. "Oh mother, mother, mother!" she sobbed. She had an impression that the Captain, beside her, if more and more friendly, was by no means unembarrassed; in a minute, however, when her eyes were clearer, he was erect in front of her, very red and nervously looking about him and whacking his leg with his stick. "Say you love her, Mr. Captain; say it, say it!" she implored.

Mr. Captain's blue eyes fixed themselves very hard. "Of *course* I love her, damn it, you know!"

At this she also jumped up; she had fished out somehow her pocket-handkerchief. "So do *I* then. I do, I do, I do!" she passionately asseverated.

"Then will you come back to her?"

Maisie, staring, stopped the tight little plug of her handkerchief on the way to her eyes. "She won't have me."

"Yes she will. She wants you."

"Back at the house—with Sir Claude?"

Again he hung fire. "No, not with him. In another place."

They stood looking at each other with an intensity unusual as between a Captain and a little girl. "She won't have me in any place."

"Oh yes she will if *I* ask her!"

Maisie's intensity continued. "Shall you be there?"

The Captain's, on the whole, did the same. "Oh yes— some day."

"Then you don't mean now?"

He broke into a quick smile. "Will you come now?—go with us for an hour?"

Maisie considered. "She would n't have me even now." She could see that he had his idea, but that her tone impressed him. That disappointed her a little, though in an instant he rang out again.

"She will if I ask her," he repeated. "I'll ask her this minute."

Maisie, turning at this, looked away to where her mother and her stepfather had stopped. At first, among the trees, nobody was visible; but the next moment she exclaimed with expression: "It's over—here he comes!"

The Captain watched the approach of her ladyship's husband, who lounged composedly over the grass, making to Maisie with his closed fingers a little movement in the air. "I've no desire to avoid him."

"Well, you must n't see him," said Maisie.

"Oh he's in no hurry himself!" Sir Claude had stopped to light another cigarette.

She was vague as to the way it was proper he should feel; but she had a sense that the Captain's remark was rather a free reflexion on it. "Oh he does n't care!" she replied.

"Does n't care for what?"

"Does n't care who you are. He told me so. Go and ask mamma," she added.

"If you can come with us? Very good. You really want me not to wait for him?"

"*Please* don't." But Sir Claude was not yet near, and the Captain had with his left hand taken hold of her right, which he familiarly, sociably swung a little. "Only first," she continued, "tell me this. Are you going to *live* with mamma?"

The immemorial note of mirth broke out at her seriousness. "One of these days."

She wondered, wholly unperturbed by his laughter. "Then where will Sir Claude be?"

"He'll have left her of course."

"Does he really intend to do that?"

"You've every opportunity to ask him."

Maisie shook her head with decision. "He won't do it. Not first."

Her "first" made the Captain laugh out again. "Oh he'll be sure to be nasty! But I've said too much to you."

"Well, you know, I'll never tell," said Maisie.

"No, it's all for yourself. Good-bye."

"Good-bye." Maisie kept his hand long enough to add: "I like you too." And then supremely: "You *do* love her?"

"My dear child—!" The Captain wanted words.

"Then don't do it only for just a little."

"A little?"

"Like all the others."

"All the others?"—he stood staring.

She pulled away her hand. "Do it always!" She bounded

to meet Sir Claude, and as she left the Captain she heard him ring out with apparent gaiety:

"Oh I'm in for it!"

As she joined Sir Claude she noted her mother in the distance move slowly off, and, glancing again at the Captain, saw him, swinging his stick, retreat in the same direction.

She had never seen Sir Claude look as he looked just then; flushed yet not excited—settled rather in an immoveable disgust and at once very sick and very hard. His conversation with her mother had clearly drawn blood, and the child's old horror came back to her, begetting the instant moral contraction of the days when her parents had looked to her to feed their love of battle. Her greatest fear for the moment, however, was that her friend would see she had been crying. The next she became aware that he had glanced at her, and it presently occurred to her that he did n't even wish to be looked at. At this she quickly removed her gaze, while he said rather curtly: "Well, who in the world *is* the fellow?"

She felt herself flooded with prudence. "Oh *I* have n't found out!" This sounded as if she meant he ought to have done so himself; but she could only face doggedly the ugliness of seeming disagreeable, as she used to face it in the hours when her father, for her blankness, called her a dirty little donkey, and her mother, for her falsity pushed her out of the room.

"Then what have you been doing all this time?"

"Oh I don't know!" It was of the essence of her method not to be silly by halves.

"Then did n't the beast say anything?" They had got down by the lake and were walking fast.

"Well, not very much."

"He did n't speak of your mother?"

"Oh yes, a little!"

"Then what I ask you, please, is *how*?" She kept silence—so long that he presently went on: "I say, you know—don't you hear me?"

At this she produced: "Well, I'm afraid I did n't attend to him very much."

Sir Claude, smoking rather hard, made no immediate rejoinder; but finally he exclaimed: "Then my dear—with such a chance—you were the perfection of a dunce!" He was so irritated—or she took him to be—that for the rest of the time they were in the Gardens he spoke no other word; and she meanwhile subtly abstained from any attempt to pacify him. That would only lead to more questions. At the gate of the Gardens he hailed a four-wheeled cab and, in silence, without meeting her eyes, put her into it, only saying "Give him *that*" as he tossed half a crown upon the seat. Even when from outside he had closed the door and told the man where to go he never took her departing look. Nothing of this kind had ever yet happened to them, but it had no power to make her love him less; so she could not only bear it, she felt as she drove away—she could rejoice in it. It brought again the sweet sense of success that, ages before, she had had at a crisis when, on the stairs, returning from her father's, she had met a fierce question of her mother's with an imbecility as deep and had in consequence been dashed by Mrs. Farange almost to the bottom.

XVII

IF for reasons of her own she could bear the sense of Sir Claude's displeasure her young endurance might have been put to a serious test. The days went by without his knocking at her father's door, and the time would have turned sadly to waste if something had n't conspicuously happened to give it a new difference. What took place was a marked change in the attitude of Mrs. Beale—a change that somehow, even in his absence, seemed to bring Sir Claude again into the house. It began practically with a conversation that occurred between them the day Maisie came home alone in the cab. Mrs. Beale had by that time returned, and she was more successful than their friend in extracting from our young lady an account of the extraordinary passage with the

Captain. She came back to it repeatedly, and on the very next day it grew distinct to the child that she was already in full possession of what at the same moment had been enacted between her ladyship and Sir Claude. This was the real origin of her final perception that though he did n't come to the house her stepmother had some rare secret for not being quite without him. This led to some rare passages with Mrs. Beale, the promptest of which had been—not on Maisie's part—a wonderful outbreak of tears. Mrs. Beale was not, as she herself said, a crying creature: she had n't cried, to Maisie's knowledge, since the lowly governess days, the grey dawn of their connexion. But she wept now with passion, professing loudly that it did her good and saying remarkable things to her charge, for whom the occasion was an equal benefit, an addition to all the fine precautionary wisdom stored away. It somehow had n't violated that wisdom, Maisie felt, for her to have told Mrs. Beale what she had not told Sir Claude, inasmuch as the greatest strain, to her sense, was between Sir Claude and Sir Claude's wife, and his wife was just what Mrs. Beale was unfortunately not. He sent his stepdaughter three days after the incident in Kensington Gardens a message as frank as it was tender, and that was how Mrs. Beale had had to bring out in a manner that seemed half an appeal, half a defiance: "Well yes, hang it— I *do* see him!"

How and when and where, however, were just what Maisie was not to know—an exclusion moreover that she never questioned in the light of a participation large enough to make him, while she shared the ample void of Mrs. Beale's rather blank independence, shine in her yearning eye like the single, the sovereign window-square of a great dim disproportioned room. As far as her father was concerned such hours had no interruption; and then it was clear between them that each was thinking of the absent and thinking the other thought, so that he was an object of conscious reference in everything they said or did. The wretched truth, Mrs. Beale had to confess, was that she had hoped against hope and that in the Regent's Park it was impossible Sir Claude

should really be in and out. Had n't they at last to look the
fact in the face?—it was too disgustingly evident that no one
after all had been squared. Well, if no one had been squared
it was because every one had been vile. No one and every one
were of course Beale and Ida, the extent of whose power to
be nasty was a thing that, to a little girl, Mrs. Beale simply
could n't give chapter and verse for. Therefore it was that
to keep going at all, as she said, that lady had to make, as
she also said, another arrangement—the arrangement in
which Maisie was included only to the point of knowing it
existed and wondering wistfully what it was. Conspicuously
at any rate it had a side that was responsible for Mrs. Beale's
sudden emotion and sudden confidence—a demonstration
this, however, of which the tearfulness was far from deter-
rent to our heroine's thought of how happy she should be if
she could only make an arrangement for herself. Mrs.
Beale's own operated, it appeared, with regularity and
frequency; for it was almost every day or two that she was
able to bring Maisie a message and to take one back. It had
been over the vision of what, as she called it, he did for her
that she broke down; and this vision was kept in a manner
before Maisie by a subsequent increase not only of the gaiety,
but literally—it seemed not presumptuous to perceive—of
the actual virtue of her friend. The friend was herself the
first to proclaim it: he had pulled her up immensely—he
had quite pulled her round. She had charming tormenting
words about him: he was her good fairy, her hidden spring—
above all he was just her "higher" conscience. That was what
had particularly come out with her startling tears: he had
made her, dear man, think ever so much better of herself.
It had been thus rather surprisingly revealed that she had
been in a way to think ill, and Maisie was glad to hear of the
corrective at the same time that she heard of the ailment.

She presently found herself supposing, and in spite of
her envy even hoping, that whenever Mrs. Beale was out of
the house Sir Claude had in some manner the satisfaction
of it. This was now of more frequent occurrence than ever
before—so much so that she would have thought of her

stepmother as almost extravagantly absent had it not been that, in the first place, her father was a superior specimen of that habit: it was the frequent remark of his present wife, as it had been, before the tribunals of their country, a prominent plea of her predecessor, that he scarce came home even to sleep. In the second place Mrs. Beale, when she *was* on the spot, had now a beautiful air of longing to make up for everything. The only shadow in such bright intervals was that, as Maisie put it to herself, she could get nothing by questions. It was in the nature of things to be none of a small child's business, even when a small child had from the first been deluded into a fear that she might be only too much initiated. Things then were in Maisie's experience so true to their nature that questions were almost always improper; but she learned on the other hand soon to recognise how at last, sometimes, patient little silences and intelligent little looks could be rewarded by delightful little glimpses. There had been years at Beale Farange's when the monosyllable "he" meant always, meant almost violently, the master; but all that was changed at a period at which Sir Claude's merits were of themselves so much in the air that it scarce took even two letters to name him. "He keeps me up splendidly—he does, my own precious," Mrs. Beale would observe to her comrade; or else she would say that the situation at the other establishment had reached a point that could scarcely be believed—the point, monstrous as it sounded, of his not having laid eyes upon her for twelve days. "She" of course at Beale Farange's had never meant any one but Ida, and there was the difference in this case that it now meant Ida with renewed intensity. Mrs. Beale—it was striking—was in a position to animadvert more and more upon her dreadfulness, the moral of all which appeared to be how abominably yet blessedly little she had to do with her husband. This flow of information came home to our two friends because, truly, Mrs. Beale had not much more to do with her own; but that was one of the reflexions that Maisie could make without allowing it to break the spell of her present sympathy. How could such a spell be anything but deep when Sir Claude's influence, operating from afar, at last really determined the resumption of his stepdaughter's studies? Mrs. Beale again took fire about them and was quite vivid for Maisie as to their being the great matter to which the dear absent one kept her up.

This was the second source—I have just alluded to the first—of the child's consciousness of something that, very hopefully, she described to herself as a new phase; and it also presented in the brightest light the fresh enthusiasm with which Mrs. Beale always reappeared and which really gave Maisie a happier sense than she had yet had of being very dear at least to two persons. That she had small remembrance at present of a third illustrates, I am afraid, a temporary oblivion of Mrs. Wix, an accident to be explained only by a state of unnatural excitement. For what was the form taken by Mrs. Beale's enthusiasm and acquiring relief in the domestic conditions still left to her but the delightful form of "reading" with her little charge on lines directly prescribed and in works profusely supplied by Sir Claude? He had got hold of an awfully good list—"mostly essays, don't you know?" Mrs. Beale had said; a word always august to Maisie, but henceforth to be softened by hazy, in fact by quite languorous edges. There was at any rate a week in which no less than nine volumes arrived, and the impression was to be gathered from Mrs. Beale that the obscure intercourse she enjoyed with Sir Claude not only involved an account and a criticism of studies, but was organised almost for the very purpose of report and consultation. It was for Maisie's education in short that, as she often repeated, she closed her door—closed it to the gentlemen who used to flock there in such numbers and whom her husband's practical desertion of her would have made it a course of the highest indelicacy to receive. Maisie was familiar from of old with the principle at least of the care that a woman, as Mrs. Beale phrased it, attractive and exposed must take of her "character," and was duly impressed with the rigour of her stepmother's scruples. There was literally no one of the other sex whom she seemed to feel at liberty to see at home, and when the child risked an enquiry about the ladies who, one by one, during her own previous period, had been made quite loudly welcome, Mrs. Beale hastened to inform her that, one by one, they had, the fiends, been found out, after all, to be awful. If she wished to know more about them she was recommended to approach her father.

Maisie had, however, at the very moment of this injunction much livelier curiosities, for the dream of lectures at an institution had at last become a reality, thanks to Sir Claude's now unbounded energy in discovering what could be done. It stood out in this connexion that when you came to look into things in a spirit of earnestness an immense deal could be done for very little more than your fare in the Underground. The institution—there was a splendid one in a part of the town but little known to the child—became, in the glow of such a spirit, a thrilling place, and the walk to it from the station through Glower Street (a pronunciation for which Mrs. Beale once laughed at her little friend) a pathway literally strewn with "subjects." Maisie imagined herself to pluck them as she went, though they thickened in the great grey rooms where the fountain of knowledge, in the form usually of a high voice that she took at first to be angry, plashed in the stillness of rows of faces thrust out like empty jugs. "It *must* do us good—it's all so hideous," Mrs. Beale had immediately declared; manifesting a purity of resolution that made these occasions quite the most harmonious of all the many on which the pair had pulled together. Maisie certainly had never, in such an association, felt so uplifted, and never above all been so carried off her feet, as at the moments of Mrs. Beale's breathlessly re-entering the house and fairly shrieking upstairs to know if they should still be in time for a lecture. Her stepdaughter, all ready from the earliest hours, almost leaped over the banister to respond, and they dashed out together in quest of learning as hard as they often dashed back to release Mrs. Beale for other preoccupations. There had been in short no bustle like these particular spasms, once they had broken out, since that last brief flurry when Mrs. Wix, blowing as if she were grooming her, "made up" for everything previously lost at her father's.

These weeks as well were too few, but they were flooded with a new emotion, part of which indeed came from the possibility that, through the long telescope of Glower Street, or perhaps between the pillars of the institution—which impressive objects were what Maisie thought most made it one—they should some day spy Sir Claude. That was what Mrs. Beale, under pressure, had said—doubtless a little impatiently: "Oh yes, oh yes, some day!" His joining them was clearly far less of a matter of course than was to have been gathered from his original profession of desire to improve in their company his own mind; and this sharpened our young lady's guess that since that occasion either something destructive had happened or something desirable had n't. Mrs. Beale had thrown but a partial light in telling her how it had turned out that nobody had been squared. Maisie wished at any rate that somebody *would* be squared. However, though in every approach to the temple of knowledge she watched in vain for Sir Claude, there was no doubt about the action of his loved image as an incentive and a recompense. When the institution was most on pillars—or, as Mrs. Beale put it, on stilts—when the subject was deepest and the lecture longest and the listeners ugliest, then it was they both felt their patron in the background would be most pleased with them.

One day, abruptly, with a glance at this background, Mrs. Beale said to her companion: "We'll go to-night to the thingumbob at Earl's Court"; an announcement putting forth its full lustre when she had made known that she referred to the great Exhibition just opened in that quarter, a collection of extraordinary foreign things in tremendous gardens, with illuminations, bands, elephants, switchbacks and side-shows, as well as crowds of people among whom they might possibly see some one they knew. Maisie flew in the same bound at the neck of her friend and at the name of Sir Claude, on which Mrs. Beale confessed that—well, yes, there was just a chance that he would be able to meet them. He never of course, in his terrible position, knew what might happen from hour to hour; but he hoped to be free and he had given Mrs. Beale the tip. "Bring her there on the quiet and I'll try to turn up"—this was clear enough on what so many weeks of privation had made of his desire to see the child: it even appeared to represent on his part a yearning as constant as her own. That in turn was just puzzling enough to make Maisie express a bewilderment. She could n't see, if they were so intensely of the same mind, why the theory on which she had come back to Mrs. Beale, the general reunion, the delightful trio, should have broken down so in fact. Mrs. Beale furthermore only gave her more to think about in saying that their disappointment was the result of his having got into his head a kind of idea.

"What kind of idea?"

"Oh goodness knows!" She spoke with an approach to asperity. "He's so awfully delicate."

"Delicate?"—that was ambiguous.

"About what he does, don't you know?" said Mrs. Beale. She fumbled. "Well, about what *we* do."

Maisie wondered. "You and me?"

"Me and *him*, silly!" cried Mrs. Beale with, this time, a real giggle.

"But you don't do any harm—*you* don't," said Maisie, wondering afresh and intending her emphasis as a decorous allusion to her parents.

"Of course we don't, you angel—that's just the ground *I* take!" her companion exultantly responded. "He says he does n't want you mixed up."

"Mixed up with what?"

"That's exactly what *I* want to know: mixed up with what, and how you are any more mixed—?" Mrs. Beale paused without ending her question. She ended after an instant in a different way. "All you can say is that it's his fancy."

The tone of this, in spite of its expressing a resignation, the fruit of weariness, that dismissed the subject, conveyed so vividly how much such a fancy was not Mrs. Beale's own that our young lady was led by the mere fact of contact to arrive at a dim apprehension of the unuttered and the unknown. The relation between her step-parents had then a mysterious residuum; this was the first time she really had reflected that except as regards herself it was not a relationship. To each other it was only what they might have happened to make it, and she gathered that this, in the event, had been something that led Sir Claude to keep away from her. Did n't he fear she would be compromised? The perception of such a scruple endeared him the more, and it flashed over her that she might simplify everything by showing him how little she made of such a danger. Had n't she lived with her eyes on it from her third year? It was the condition most frequently discussed at the Faranges', where the word was always in the air and where at the age of five, amid rounds of applause, she could gabble it off. She knew as well in short that a person could be compromised as that a person could be slapped with a hair-brush or left alone in the dark, and it was equally familiar to her that each of these ordeals was in general held to have too little effect. But the first thing was to make absolutely sure of Mrs. Beale. This was done by saying to her thoughtfully: "Well, if you don't mind—and you really don't, do you?"

Mrs. Beale, with a dawn of amusement, considered. "Mixing you up? Not a bit. For what does it mean?"

"Whatever it means I don't in the least mind *being* mixed. Therefore if you don't and I don't," Maisie concluded, "don't you think that when I see him this evening I had better just tell him we don't and ask him why in the world *he* should?"

XVIII

THE child, however, was not destined to enjoy much of Sir Claude at the "thingumbob," which took for them a very different turn indeed. On the spot Mrs. Beale, with hilarity, had urged her to the course proposed; but later, at the

Exhibition, she withdrew this allowance, mentioning as a result of second thoughts that when a man was so sensitive anything at all frisky usually made him worse. It would have been hard indeed for Sir Claude to be "worse," Maisie felt, as, in the gardens and the crowd, when the first dazzle had dropped, she looked for him in vain up and down. They had all their time, the couple, for frugal wistful wandering: they had partaken together at home of the light vague meal—Maisie's name for it was a "jam-supper"—to which they were reduced when Mr. Farange sought his pleasure abroad. It was abroad now entirely that Mr. Farange pursued this ideal, and it was the actual impression of his daughter, derived from his wife, that he had three days before joined a friend's yacht at Cowes.

The place was full of side-shows, to which Mrs. Beale could introduce the little girl only, alas, by revealing to her so attractive, so enthralling a name: the side-shows, each time, were sixpence apiece, and the fond allegiance enjoyed by the elder of our pair had been established from the earliest time in spite of a paucity of sixpences. Small coin dropped from her as half-heartedly as answers from bad children to lessons that had not been looked at. Maisie passed more slowly the great painted posters, pressing with a linked arm closer to her friend's pocket, where she hoped for the audible chink of a shilling. But the upshot of this was but to deepen her yearning: if Sir Claude would only at last come the shillings would begin to ring. The companions paused, for want of one, before the Flowers of the Forest, a large present-ment of bright brown ladies—they were brown all over—in a medium suggestive of tropical luxuriance, and there Maisie dolorously expressed her belief that he would never come at all. Mrs. Beale hereupon, though discernibly disappointed, reminded her that he had not been promised as a certainty—a remark that caused the child to gaze at the Flowers through a blur in which they became more magnificent, yet oddly more confused, and by which moreover confusion was imparted to the aspect of a gentleman who at that moment, in the company of a lady, came out of the brilliant booth. The lady was so brown that Maisie at first took her for one of the Flowers; but during the few seconds that this required —a few seconds in which she had also desolately given up Sir Claude—she heard Mrs. Beale's voice, behind her, gather both wonder and pain into a single sharp little cry.

"Of all the wickedness—*Beale*!"

He had already, without distinguishing them in the mass of strollers, turned another way—it seemed at the brown lady's suggestion. Her course was marked, over heads and shoulders, by an upright scarlet plume, as to the ownership of which Maisie was instantly eager. "Who is she?—who is she?"

But Mrs. Beale for a moment only looked after them. "The liar—the liar!"

Maisie considered. "Because he's not—where one thought?" That was also, a month ago in Kensington Gardens, where her mother had not been. "Perhaps he has come back," she was quick to contribute.

"He never went—the hound!"

That, according to Sir Claude, had been also what her mother had not done, and Maisie could only have a sense of something that in a maturer mind would be called the way history repeats itself. "Who *is* she?" she asked again.

Mrs. Beale, fixed to the spot, seemed lost in the vision of an opportunity missed. "If he had only seen me!"—it came from between her teeth. "She's a brand-new one. But he must have been with her since Tuesday."

Maisie took it in. "She's almost black," she then reported.

"They're always hideous," said Mrs. Beale.

This was a remark on which the child had again to reflect. "Oh not his *wives*!" she remonstrantly exclaimed. The words at another moment would probably have set her friend "off," but Mrs. Beale was now, in her instant vigilance, too immensely "on." "Did you ever in your life see such a feather?" Maisie presently continued.

This decoration appeared to have paused at some distance, and in spite of intervening groups they could both look at it. "Oh that's the way they dress—the vulgarest of the vulgar!"

"They're coming back—they'll see us!" Maisie the next moment cried; and while her companion answered that this was exactly what she wanted and the child returned "Here they are—here they are!" the unconscious subjects of so much attention, with a change of mind about their direction, quickly retraced their steps and precipitated themselves upon their critics. Their unconsciousness gave Mrs. Beale time to leap, under her breath, to a recognition which Maisie caught.

"It must be Mrs. Cuddon!"

Maisie looked at Mrs. Cuddon hard—her lips even echoed the name. What followed was extraordinarily rapid—a minute of livelier battle than had ever yet, in so short a span at least, been waged round our heroine. The muffled shock—lest people should notice—was violent, and it was only for her later thought that the steps fell into their order, the steps through which, in a bewilderment not so much of sound as of silence, she had come to find herself, too soon for comprehension and too strangely for fear, at the door of the Exhibition with her father. He thrust her into a hansom and got in after her, and then it was—as she drove along with him—that she recovered a little what had happened. Face to face with them in the gardens he had seen them, and there had been a moment of checked concussion during which, in a glare of black eyes and a toss of red plumage, Mrs. Cuddon had recognised them, ejaculated and vanished. There had been another moment at which she became aware of Sir Claude, also poised there in surprise, but out of her father's view, as if he had been warned off at the very moment of reaching them. It fell into its place with all the rest that she had heard Mrs. Beale say to her father, but whether low or loud was now lost to her, something about his having this time a new one; on which he had growled something indistinct but apparently in the tone and of the sort that the child, from her earliest years, had associated with hearing somebody retort to somebody that somebody was "another,"

"Oh I stick to the old!" Mrs. Beale had then quite loudly pronounced; and her accent, even as the cab got away, was still in the air, Maisie's effective companion having spoken no other word from the moment of whisking her off—none at least save the indistinguishable address which, over the top of the hansom and poised on the step, he had given the driver. Reconstructing these things later Maisie theorised that she at this point would have put a question to him had not the silence into which he charmed her or scared her— she could scarcely tell which—come from his suddenly making her feel his arm about her, feel, as he drew her close, that he was agitated in a way he had never yet shown her. It struck her he trembled, trembled too much to speak, and this had the effect of making her, with an emotion which, though it had begun to throb in an instant, was by no means all dread, conform to his portentous hush. The act of possession that his pressure in a manner advertised came back to her after the longest of the long intermissions that had ever let anything come back. They drove and drove, and he kept her close; she stared straight before her, holding her breath, watching one dark street succeed another and strangely conscious that what it all meant was somehow that papa was less to be left out of everything than she had supposed. It took her but a minute to surrender to this discovery, which, in the form of his present embrace, suggested a purpose in him prodigiously reaffirmed and with that a confused confidence. She neither knew exactly what he had done nor what he was doing; she could only, altogether impressed and rather proud, vibrate with the sense that he had jumped up to do something and that she had as quickly become a part of it. It was a part of it too that here they were at a house that seemed not large, but in the fresh white front of which the street-lamp showed a smartness of flower-boxes. The child had been in thousands of stories—all Mrs. Wix's and her own, to say nothing of the richest romances of French Elise— but she had never been in such a story as this. By the time he had helped her out of the cab, which drove away, and she heard in the door of the house the prompt little click of his key, the Arabian Nights had quite closed round her.

From this minute that pitch of the wondrous was in everything, particularly in such an instant "Open Sesame" and in the departure of the cab, a rattling void filled with relinquished step-parents; it was, with the vividness, the almost blinding whiteness of the light that sprang responsive to papa's quick touch of a little brass knob on the wall, in a place that, at the top of a short soft staircase, struck her as the most beautiful she had ever seen in her life. The next thing she perceived it to be was the drawing-room of a lady—oh of a lady, she could see in a moment, and not of a gentleman, not even of one like papa himself or even like Sir Claude—whose things were as much prettier than mamma's as it had always had to be confessed that mamma's were prettier than Mrs. Beale's. In the middle of the small bright room and the presence of more curtains and cushions, more pictures and mirrors, more palm-trees drooping over brocaded and gilded nooks, more little silver boxes scattered over little crooked tables and little oval miniatures hooked upon velvet screens than Mrs. Beale and her ladyship together could, in an unnatural alliance, have dreamed of mustering, the child became aware, with a sharp foretaste of compassion, of something that was strangely like a relegation to obscurity of each of those women of taste. It was a stranger operation still that her father should on the spot be presented to her as quite advantageously and even grandly at home in the dazzling scene and himself by so much the more separated from scenes inferior to it. She spent with him in it, while explanations continued to hang back, twenty minutes that, in their sudden drop of danger, affected her, though there were neither buns nor ginger-beer, like an extemporised expensive treat.

"Is she very rich?" He had begun to strike her as almost embarrassed, so shy that he might have found himself with a young lady with whom he had little in common. She was literally moved by this apprehension to offer him some tactful relief.

Beale Farange stood and smiled at his young lady, his back to the fanciful fireplace, his light overcoat—the very lightest in London—wide open, and his wonderful lustrous beard completely concealing the expanse of his shirt-front. It pleased her more than ever to think that papa was handsome and, though as high aloft as mamma and almost, in his specially florid evening-dress, as splendid, of a beauty somehow less belligerent, less terrible. "The Countess? Why do you ask me that?"

Maisie's eyes opened wider. "Is she a Countess?"

He seemed to treat her wonder as a positive tribute. "Oh yes, my dear, but it is n't an English title."

Her manner appreciated this. "Is it a French one?"

"No, nor French either. It's American."

She conversed agreeably. "Ah then of course she must be rich." She took in such a combination of nationality and rank. "I never saw anything so lovely."

"Did you have a sight of her?" Beale asked.

"At the Exhibition?" Maisie smiled. "She was gone too quick."

Her father laughed. "She did slope!" She had feared he would say something about Mrs. Beale and Sir Claude, yet the way he spared them made her rather uneasy too. All he risked was, the next minute, "She has a horror of vulgar scenes."

This was something she need n't take up; she could still continue bland. "But where do you suppose she went?"

"Oh I thought she'd have taken a cab and have been here by this time. But she'll turn up all right."

"I'm sure I *hope* she will," Maisie said; she spoke with an earnestness begotten of the impression of all the beauty about them, to which, in person, the Countess might make further contribution. "We came awfully fast," she added.

Her father again laughed loud. "Yes, my dear, I made you step out!" He waited an instant, then pursued: "I want her to see you."

Maisie, at this, rejoiced in the attention that, for their evening out, Mrs. Beale, even to the extent of personally "doing up" her old hat, had given her appearance. Meanwhile her father went on: "You'll like her awfully."

"Oh I'm sure I shall!" After which, either from the effect of having said so much or from that of a sudden glimpse of the impossibility of saying more, she felt an embarrassment and sought refuge in a minor branch of the subject. "I thought she was Mrs. Cuddon."

Beale's gaiety rather increased than diminished. "You mean my wife did? My dear child, my wife's a damned fool!" He had the oddest air of speaking of his wife as of a person whom she might scarcely have known, so that the refuge of her scruple didn't prove particularly happy. Beale on the other hand appeared after an instant himself to feel a scruple. "What I mean is, to speak seriously, that she does n't really know anything about anything." He paused, following the child's charmed eyes and tentative step or two as they brought her nearer to the pretty things on one of the tables. "She thinks she has good things, don't you know!" He quite jeered at Mrs. Beale's delusion.

Maisie felt she must confess that it *was* one; everything she had missed at the side-shows was made up to her by the Countess's luxuries. "Yes," she considered; "she does think that."

There was again a dryness in the way Beale replied that it did n't matter what she thought; but there was an increasing sweetness for his daughter in being with him so long without his doing anything worse. The whole hour of course was to remain with her, for days and weeks, ineffaceably illumined and confirmed; by the end of which she was able to read into it a hundred things that had been at the moment mere miraculous pleasantness. What they at the moment came to was simply that her companion was still in a good deal of a flutter, yet wished not to show it, and that just in proportion as he succeeded in this attempt he was able to encourage her to regard him as kind. He moved about the room after a little, showed her things, spoke to her as a person of taste, told her the name, which she remembered, of the famous French lady represented in one of the miniatures, and remarked, as if he had caught her wistful over a trinket or a trailing stuff, that he made no doubt the Countess, on coming in, would give her something jolly. He spied a pink satin box with a looking-glass let into the cover, which he raised, with a quick facetious flourish, to offer her the privilege of six rows of chocolate bonbons, cutting out thereby Sir Claude, who had never gone beyond four rows. "I can do what I like with these," he said, "for I don't mind telling you I gave 'em to her myself." The Countess had evidently appreciated the gift; there were numerous gaps, a ravage now quite unchecked, in the array. Even while they waited together Maisie had her sense, which was the mark of what their separation had become, of her having grown for him, since the last time he had, as it were, noticed her, and by increase of year and of inches if by nothing else, much more of a little person to reckon with. Yes, this was a part of the positive awkwardness that he carried off by being almost foolishly tender. There was a passage during which, on a yellow silk sofa under one of the palms, he had her on his knee, stroking her hair, playfully holding her off while he showed his shining fangs and let her, with a vague affectionate helpless pointless "Dear old girl, dear little daughter," inhale the fragrance of his cherished beard. She must have been sorry for him, she afterwards knew, so well could she privately follow his difficulty in being specific to her about anything. She had such possibilities of vibration, of response, that it needed nothing more than this to make up to her in fact for omissions. The tears came into her eyes again as they had done when in the Park that day the Captain told her so "splendidly" that her mother was good. What was this but splendid too—this still directer goodness of her father and this unexampled shining solitude with him, out of which everything had dropped but that he was papa and that he was magnificent? It did n't spoil it that she finally felt he must have, as he became restless, some purpose he did n't quite see his way to bring out, for in the freshness of their recovered fellowship she would have lent herself gleefully to his suggesting, or even to his pretending, that their relations were easy and graceful. There was something in him that seemed, and quite touchingly, to ask her to help him to pretend—pretend he knew enough about her life and her education, her means of subsistence and her view of himself, to give the questions he could n't put her a natural domestic tone. She would have pretended with ecstasy if he could only have given her the cue. She waited for it while, between his big teeth, he breathed the sighs she did n't know to be stupid. And as if, though he was so stupid all through, he had let the friendly suffusion of her eyes yet tell him she was ready for anything, he floundered about, wondering what the devil he could lay hold of.

XIX

When he had lighted a cigarette and begun to smoke in her face it was as if he had struck with the match the note of some queer clumsy ferment of old professions, old scandals, old duties, a dim perception of what he possessed in her and what, if everything had only—damn it!—been totally different, she might still be able to give him. What she was able to give him, however, as his blinking eyes seemed to make out through the smoke, would be simply what he should be able to get from her. To give something, to give here on the spot, was all her own desire. Among the old things that came back was her little instinct of keeping the peace; it made her wonder more sharply what particular thing she could do or not do, what particular word she could speak or not speak, what particular line she could take or not take, that might for every one, even for the Countess, give a better turn to the crisis. She was ready, in this interest, for an immense surrender, a surrender of everything but Sir Claude, of everything but Mrs. Beale. The immensity did n't include *them*; but if he had an idea at the back of his head she had also one in a recess as deep, and for a time, while they sat together, there was an extraordinary mute passage between her vision of this vision of his, his vision of her vision, and her vision of his vision of her vision. What there was no effective record of indeed was the small strange pathos on the child's part of an innocence so saturated with knowledge and so directed to diplomacy. What, further, Beale finally laid hold of while he masked again with his fine presence half the flounces of the fireplace was: "Do you know, my dear, I shall soon be off to America?" It struck his daughter both as a short cut and as the way he would n't have said it to his wife. But his wife figured with a bright superficial assurance in her response.

"Do you mean with Mrs. Beale?"

Her father looked at her hard. "Don't be a little ass!"

Her silence appeared to represent a concentrated effort not to be. "Then with the Countess?"

"With her or without her, my dear; that concerns only your poor daddy. She has big interests over there, and she wants me to take a look at them."

Maisie threw herself into them. "Will that take very long?"

"Yes; they're in such a muddle—it may take months. Now what I want to hear, you know, is whether you'd like to come along?"

Planted once more before him in the middle of the room she felt herself turning white. "I?" she gasped, yet feeling as soon as she had spoken that such a note of dismay was not altogether pretty. She felt it still more while her father replied, with a shake of his legs, a toss of his cigarette-ash and a fidgety look—he was for ever taking one—all the length of his waistcoat and trousers, that she need n't be quite so disgusted. It helped her in a few seconds to appear more as he would like her that she saw, in the lovely light of the Countess's splendour, exactly, however she appeared, the right answer to make. "Dear papa, I'll go with you anywhere."

He turned his back to her and stood with his nose at the glass of the chimneypiece while he brushed specks of ash out of his beard. Then he abruptly said: "Do you know anything about your brute of a mother?"

It was just of her brute of a mother that the manner of the

question in a remarkable degree reminded her: it had the free flight of one of Ida's fine bridgings of space. With the sense of this was kindled for Maisie at the same time an inspiration. "Oh yes, I know everything!" and she became so radiant that her father, seeing it in the mirror, turned back to her and presently, on the sofa, had her at his knee again and was again particularly affecting. Maisie's inspiration instructed her, pressingly, that the more she should be able to say about mamma the less she would be called upon to speak of her step-parents. She kept hoping the Countess would come in before her power to protect them was exhausted; and it was now, in closer quarters with her companion, that the idea at the back of her head shifted its place to her lips. She told him she had met her mother in the Park with a gentleman who, while Sir Claude had strolled with her ladyship, had been kind and had sat and talked to her; narrating the scene with a remembrance of her pledge of secrecy to the Captain quite brushed away by the joy of seeing Beale listen without profane interruption. It was almost an amazement, but it was indeed all a joy, thus to be able to guess that papa was at last quite tired of his anger—of his anger at any rate about mamma. He was only bored with her now. That made it, however, the more imperative that his spent displeasure should n't be blown out again. It charmed the child to see how much she could interest him; and the charm remained even when, after asking her a dozen questions, he observed musingly and a little obscurely: "Yes, damned if she won't!" For in this too there was a detachment, a wise weariness that made her feel safe. She had had to mention Sir Claude, though she mentioned him as little as possible and Beale only appeared to look quite over his head. It pieced itself together for her that this was the mildness of general indifference, a source of profit so great for herself personally that if the Countess was the author of it she was prepared literally to hug the Countess. She betrayed that eagerness by a restless question about her, to which her father replied:

"Oh she has a head on her shoulders. I 'll back her to get

out of anything!" He looked at Maisie quite as if he could trace the connexion between her enquiry and the impatience of her gratitude. "Do you mean to say you'd really come with me?"

She felt as if he were now looking at her very hard indeed, and also as if she had grown ever so much older. "I'll do anything in the world you ask me, papa."

He gave again, with a laugh and with his legs apart, his proprietary glance at his waistcoat and trousers. "That's a way, my dear, of saying 'No, thank you!' You know you don't want to go the least little mite. You can't humbug *me*!" Beale Farange laid down. "I don't want to bully you—I never bullied you in my life; but I make you the offer, and it's to take or to leave. Your mother will never again have any more to do with you than if you were a kitchenmaid she had turned out for going wrong. Therefore of course I'm your natural protector and you've a right to get everything out of me you can. Now's your chance, you know—you won't be half-clever if you don't. You can't say I don't put it before you—you can't say I ain't kind to you or that I don't play fair. Mind you never say that, you know—it *would* bring me down on you. I know what's proper. I'll take you again, just as I *have* taken you again and again. And I'm much obliged to you for making up such a face."

She was conscious enough that her face indeed could n't please him if it showed any sign—just as she hoped it did n't—of her sharp impression of what he now really wanted to do. Was n't he trying to turn the tables on her, embarrass her somehow into admitting that what would really suit her little book would be, after doing so much for good manners, to leave her wholly at liberty to arrange for herself? She began to be nervous again: it rolled over her that this was their parting, their parting for ever, and that he had brought her there for so many caresses only because it was important such an occasion should look better for him than any other. For her to spoil it by the note of discord would certainly give him ground for complaint; and the child was momentarily bewildered between her alternatives of agreeing with him about her wanting to get rid of him and displeasing him by pretending to stick to him. So she found for the moment no solution but to murmur very helplessly: "Oh papa—oh papa!"

"I know what you're up to—don't tell *me*!" After which he came straight over and, in the most inconsequent way in the world, clasped her in his arms a moment and rubbed his beard against her cheek. Then she understood as well as if he had spoken it that what he wanted, hang it, was that she should let him off with all the honours—with all the appearance of virtue and sacrifice on his side. It was exactly as if he had broken out to her: "I say, you little booby, help me to be irreproachable, to be noble, and yet to have none of the beastly bore of it. There's only impropriety enough for one of us; so *you* must take it all. *Repudiate* your dear old daddy—in the face, mind you, of his tender supplications. He can't be rough with you—it is n't in his nature: therefore you'll have successfully chucked him because he was too generous to be as firm with you, poor man, as was, after all, his duty." This was what he communicated in a series of tremendous pats on the back; that portion of her person had never been so thumped since Moddle thumped her when she choked. After a moment he gave her the further impression of having become sure enough of her to be able very gracefully to say out: "You know your mother loathes you, loathes you simply. And I've been thinking over your precious man—the fellow you told me about."

"Well," Maisie replied with competence, "I'm sure of *him*."

Her father was vague for an instant. "Do you mean sure of his liking you?"

"Oh no; of his liking *her*!"

Beale had a return of gaiety. "There's no accounting for tastes! It's what they all say, you know."

"I don't care—I'm sure of him!" Maisie repeated.

"Sure, you mean, that she'll bolt?"

Maisie knew all about bolting, but, decidedly, she *was* older, and there was something in her that could wince at the way her father made the ugly word—ugly enough at best—sound flat and low. It prompted her to amend his allusion, which she did by saying: "I don't know what she'll do. But she'll be happy."

"Let us hope so," said Beale—almost as for edification. "The more happy she is at any rate the less she'll want you about. That's why I press you," he agreeably pursued, "to consider this handsome offer—I mean seriously, you know—of your sole surviving parent." Their eyes, at this, met again in a long and extraordinary communion which terminated in his ejaculating: "Ah you little scoundrel!" She took it from him in the manner it seemed to her he would like best and with a success that encouraged him to go on: "You *are* a deep little devil!" Her silence, ticking like a watch, acknowledged even this, in confirmation of which he finally brought out: "You've settled it with the other pair!"

"Well, what if I have?" She sounded to herself most bold.

Her father, quite as in the old days, broke into a peal. "Why, don't you know they're awful?"

She grew bolder still. "I don't care—not a bit!"

"But they're probably the worst people in the world and the very greatest criminals," Beale pleasantly urged. "I'm not the man, my dear, not to let you know it."

"Well, it doesn't prevent them from loving me. They love me tremendously." Maisie turned crimson to hear herself.

Her companion fumbled; almost any one—let alone a daughter—would have seen how conscientious he wanted to be. "I dare say. But do you know why?" She braved his eyes and he added: "You're a jolly good pretext."

"For what?" Maisie asked.

"Why, for their game. I needn't tell you what that is."

The child reflected. "Well then that's all the more reason."

"Reason for what, pray?"

"For their being kind to me."

"And for your keeping in with them?" Beale roared again; it was as if his spirits rose and rose. "Do you realise, pray, that in saying that you're a monster?"

She turned it over. "A monster?"

"They've *made* one of you. Upon my honour it's quite awful. It shows the kind of people they are. Don't you understand," Beale pursued, "that when they've made you as horrid as they can—as horrid as themselves—they'll just simply chuck you?"

She had at this a flicker of passion. "They *won't* chuck me!"

"I beg your pardon," her father courteously insisted; "it's my duty to put it before you. I should n't forgive myself if I did n't point out to you that they'll cease to require you." He spoke as if with an appeal to her intelligence that she must be ashamed not adequately to meet, and this gave a real distinction to his superior delicacy.

It cleared the case as he had wished. "Cease to require me because they won't care?" She paused with that sketch of her idea.

"*Of course* Sir Claude won't care if his wife bolts. That's his game. It will suit him down to the ground."

This was a proposition Maisie could perfectly embrace, but it still left a loophole for triumph. She turned it well over. "You mean if mamma does n't come back ever at all?" The composure with which her face was presented to that prospect would have shown a spectator the long road she had travelled. "Well, but that won't put Mrs. Beale—"

"In the same comfortable position—?" Beale took her up with relish; he had sprung to his feet again, shaking his legs and looking at his shoes. "Right you are darling! Something more will be wanted for Mrs. Beale." He just paused, then he added: "But she may not have long to wait for it."

Maisie also for a minute looked at his shoes, though they were not the pair she most admired, the laced yellow "uppers" and patent-leather complement. At last, with a question, she raised her eyes. "Are n't you coming back?"

Once more he hung fire; after which he gave a small laugh that in the oddest way in the world reminded her of the unique sounds she had heard emitted by Mrs. Wix. "It may strike you as extraordinary that I should make you such an admission; and in point of fact you're not to understand that

I do. But we'll put it that way to help your decision. The point is that that's the way my wife will presently be sure to put it. You'll hear her shrieking that she's deserted, so that she may just pile up her wrongs. She'll be as free as she likes then—as free, you see, as your mother's muff of a husband. They won't have anything more to consider and they'll just put you into the street. Do I understand," Beale enquired, "that, in the face of what I press on you, you still prefer to take the risk of that?" It was the most wonderful appeal any gentleman had ever addressed to his daughter, and it had placed Maisie in the middle of the room again while her father moved slowly about her with his hands in his pockets and something in his step that seemed, more than anything else he had done, to show the habit of the place. She turned her fevered little eyes over his friend's brightnesses, as if, on her own side, to press for some help in a quandary unexampled. As if also the pressure reached him he after an instant stopped short, completing the prodigy of his attitude and the pride of his loyalty by a supreme formulation of the general inducement. "You've an eye, love! Yes, there's money. No end of money."

This affected her at first in the manner of some great flashing dazzle in one of the pantomimes to which Sir Claude had taken her: she saw nothing in it but what it directly conveyed. "And shall I never, never see you again—?"

"If I do go to America?" Beale brought it out like a man. "Never, never, never!"

Hereupon, with the utmost absurdity, she broke down; everything gave way, everything but the horror of hearing herself definitely utter such an ugliness as the acceptance of that. So she only stiffened herself and said: "Then I can't give you up."

She held him some seconds looking at her, showing her a strained grimace, a perfect parade of all his teeth, in which it seemed to her she could read the disgust he did n't quite like to express at this departure from the pliability she had practically promised. But before she could attenuate in any way the crudity of her collapse he gave an impatient jerk which took him to the window. She heard a vehicle stop; Beale looked out; then he freshly faced her. He still said nothing, but she knew the Countess had come back. There was a silence again between them, but with a different shade of embarrassment from that of their united arrival; and it was still without speaking that, abruptly repeating one of the embraces of which he had already been so prodigal, he whisked her back to the lemon sofa just before the door of the room was thrown open. It was thus in renewed and intimate union with him that she was presented to a person whom she instantly recognised as the brown lady.

The brown lady looked almost as astonished, though not quite as alarmed, as when, at the Exhibition, she had gasped in the face of Mrs. Beale. Maisie in truth almost gasped in her own; this was with the fuller perception that she was brown indeed. She literally struck the child more as an animal than as a "real" lady; she might have been a clever frizzled poodle in a frill or a dreadful human monkey in a spangled petticoat. She had a nose that was far too big and eyes that were far too small and a moustache that was, well, not so happy a feature as Sir Claude's. Beale jumped up to her; while, to the child's astonishment, though as if in a quick intensity of thought, the Countess advanced as gaily as if, for many a day, nothing awkward had happened for any one. Maisie, in spite of a large acquaintance with the phenomenon, had never seen it so promptly established that nothing awkward was to be mentioned. The next minute the Countess had kissed her and exclaimed to Beale with bright tender reproach: "Why, you never told me *half*! My dear child," she cried, "it was awfully nice of you to come!"

"But she has n't come—she won't come!" Beale answered. "I 've put it to her how much you 'd like it, but she declines to have anything to do with us."

The Countess stood smiling, and after an instant that was mainly taken up with the shock of her weird aspect Maisie felt herself reminded of another smile, which was not ugly, though also interested—the kind light thrown, that day in the Park, from the clean fair face of the Captain. Papa's

Captain—yes—was the Countess; but she was n't nearly so
nice as the other: it all came back, doubtless, to Maisie's
minor appreciation of ladies. "Should n't you like me," said
this one endearingly, "to take you to Spa?"

"To Spa?" The child repeated the name to gain time,
not to show how the Countess brought back to her a dim
remembrance of a strange woman with a horrid face who
once, years before, in an omnibus, bending to her from an
opposite seat, had suddenly produced an orange and murmured "Little dearie, won't you have it?" She had felt then,
for some reason, a small silly terror, though afterwards
conscious that her interlocutress, unfortunately hideous,
had particularly meant to be kind. This was also what the
Countess meant; yet the few words she had uttered and the
smile with which she had uttered them immediately cleared
everything up. Oh no, she wanted to go nowhere with *her*,
for her presence had already, in a few seconds, dissipated the
happy impression of the room and put an end to the pleasure
briefly taken in Beale's command of such elegance. There
was no command of elegance in his having exposed her to the
approach of the short fat wheedling whiskered person in
whom she had now to recognise the only figure wholly without attraction involved in any of the intimate connexions
her immediate circle had witnessed the growth of. She was
abashed meanwhile, however, at having appeared to weigh
in the balance the place to which she had been invited; and
she added as quickly as possible: "It is n't to America then?"
The Countess, at this, looked sharply at Beale, and Beale,
airily enough, asked what the deuce it mattered when she had
already given him to understand she wanted to have nothing
to do with them. There followed between her companions
a passage of which the sense was drowned for her in the
deepening inward hum of her mere desire to get off; though
she was able to guess later on that her father must have put
it to his friend that it was no use talking, that she was an
obstinate little pig and that, besides, she was really old
enough to choose for herself. It glimmered back to her
indeed that she must have failed quite dreadfully to seem

ideally other than rude, inasmuch as before she knew it she
had visibly given the impression that if they did n't allow
her to go home she should cry. Oh if there had ever been
a thing to cry about it was being so consciously and gawkily
below the handsomest offers any one could ever have
received. The great pain of the thing was that she could see
the Countess liked her enough to wish to be liked in return,
and it was from the idea of a return she sought utterly to flee.
It was the idea of a return that after a confusion of loud words
had broken out between the others brought to her lips with
the tremor preceding disaster: "Can't I, please, be sent home
in a cab?" Yes, the Countess wanted her and the Countess
was wounded and chilled, and she could n't help it, and it
was all the more dreadful because it only made the Countess
more coaxing and more impossible. The only thing that
sustained either of them perhaps till the cab came—Maisie
presently saw it would come—was its being in the air somehow that Beale had done what he wanted. He went out to look
for a conveyance; the servants, he said, had gone to bed, but
she should n't be kept beyond her time. The Countess left
the room with him, and, alone in the possession of it, Maisie
hoped she would n't come back. It was all the effect of her
face—the child simply could n't look at it and meet its
expression halfway. All in a moment too that queer expression had leaped into the lovely things—all in a moment she
had had to accept her father as liking some one whom she
was sure neither her mother, nor Mrs. Beale, nor Mrs. Wix,
nor Sir Claude, nor the Captain, nor even Mr. Perriam and
Lord Eric could possibly have liked. Three minutes later,
downstairs, with the cab at the door, it was perhaps as a final
confession of not having much to boast of that, on taking
leave of her, he managed to press her to his bosom without
her seeing his face. For herself she was so eager to go that
their parting reminded her of nothing, not even of a single
one of all the "nevers" that above, as the penalty of not
cleaving to him, he had attached to the question of their
meeting again. There was something in the Countess that
falsified everything, even the great interests in America and

yet more the first flush of that superiority to Mrs. Beale and to mamma which had been expressed in Sèvres sets and silver boxes. These were still there, but perhaps there were no great interests in America. Mamma had known an American who was not a bit like this one. She was not, however, of noble rank; her name was only Mrs. Tucker. Maisie's detachment would none the less have been more complete if she had not suddenly had to exclaim: "Oh dear, I have n't any money!"

Her father's teeth, at this, were such a picture of appetite without action as to be a match for any plea of poverty. "Make your stepmother pay."

"Stepmothers *don't* pay!" cried the Countess. "No stepmother ever paid in her life!" The next moment they were in the street together, and the next the child was in the cab, with the Countess, on the pavement, but close to her, quickly taking money from a purse whisked out of a pocket. Her father had vanished and there was even yet nothing in that to reawaken the pang of loss. "Here's money," said the brown lady: "go!" The sound was commanding: the cab rattled off. Maisie sat there with her hand full of coin. All that for a cab? As they passed a street-lamp she bent to see how much. What she saw was a cluster of sovereigns. There *must* then have been great interests in America. It was still at any rate the Arabian Nights.

XX

THE money was far too much even for a fee in a fairy-tale, and in the absence of Mrs. Beale, who, though the hour was now late, had not yet returned to the Regent's Park, Susan Ash, in the hall, as loud as Maisie was low and as bold as she was bland, produced, on the exhibition offered under the dim vigil of the lamp that made the place a contrast to the child's recent scene of light, the half-crown that an unsophisticated cabman could pronounce to be the least he would take. It was apparently long before Mrs. Beale would arrive, and in the interval Maisie had been induced by the prompt Susan not only to go to bed like a darling dear, but, in still richer expression of that character, to devote to the repayment of obligations general as well as particular one of the sovereigns in the ordered array that, on the dressing-table upstairs, was naturally not less dazzling to a lone orphan of a housemaid than to the subject of the manœuvres of a quartette. This subject went to sleep with her property gathered into a knotted handkerchief, the largest that could be produced and lodged under her pillow; but the explanations that on the morrow were inevitably more complete with Mrs. Beale than they had been with her humble friend found their climax in a surrender also more becomingly free. There were explanations indeed that Mrs. Beale had to give as well as to ask, and the most striking of these was to the effect that it was dreadful for a little girl to take money from a woman who was simply the vilest of their sex. The sovereigns were examined with some attention, the result of which, however, was to make the author of that statement desire to know what, if one really went into the matter, they could be called but the wages of sin. Her companion went into it merely so far as the question of what then they were to do with them; on which Mrs. Beale, who had by this time put them into her pocket, replied with dignity and with her hand on the place: "We're to send them back on the spot!" Susan, the child soon afterwards learnt, had been invited to contribute to this act of restitution her one appropriated coin; but a closer clutch of the treasure showed in her private assurance to Maisie that there was a limit to the way she could be "done." Maisie had been open with Mrs. Beale about the whole of last night's transaction; but she now found herself on the part of their indignant inferior a recipient of remarks that were so many ringing tokens of that lady's own suppressions. One of these bore upon the extraordinary hour—it was three in the morning if she really wanted to know—at which Mrs. Beale had re-entered the house; another, in accents as to which Maisie's criticism was still intensely tacit, characterised her appeal as such a "gime," such a "shime," as one had never had to put up with; a third treated with some vigour the question of the enormous sums due belowstairs, in every department, for gratuitous labour and wasted zeal. Our young lady's consciousness was indeed mainly filled for several days with the apprehension created by the too slow subsidence of her attendant's sense of wrong. These days would become terrific like the Revolutions she had learnt by heart in Histories if an outbreak in the kitchen should crown them; and to promote that prospect she had through Susan's eyes more than one glimpse of the way in which Revolutions are prepared. To listen to Susan was to gather that the spark applied to the inflammables and already causing them to crackle would prove to have been the circumstance of one's being called a horrid low thief for refusing to part with one's own.

The redeeming point of this tension was, on the fifth day, that it actually appeared to have had to do with a breathless perception in our heroine's breast that scarcely more as the centre of Sir Claude's than as that of Susan's energies she had soon after breakfast been conveyed from London to Folkestone and established at a lovely hotel. These agents, before her wondering eyes, had combined to carry through the adventure and to give it the air of having owed its success to that fact that Mrs. Beale had, as Susan said, but just stepped out. When Sir Claude, watch in hand, had met this fact with the exclamation "Then pack Miss Farange and come off with us!" there had ensued on the stairs a series of gymnastics of a nature to bring Miss Farange's heart into Miss Farange's mouth. She sat with Sir Claude in a four-wheeler while he still held his watch; held it longer than any doctor who had ever felt her pulse; long enough to give her a vision of something like the ecstasy of neglecting such an opportunity to show impatience. The ecstasy had begun in the school-room and over the Berceuse, quite in the manner of the same foretaste on the day, a little while back, when Susan had panted up and she herself, after the hint about the duchess, had sailed down; for what harm then had there been in drops and disappointments if she could still have, even only a moment, the sensation of such a name "brought up"? It had remained with her that her father had foretold her she would some day be in the street, but it clearly would n't be this day, and she felt justified of the preference betrayed to that parent as soon as her visitor had set Susan in motion and laid his hand, while she waited with him, kindly on her own. This was what the Captain, in Kensington Gardens, had done; her present situation reminded her a little of that one and renewed the dim wonder of the fashion after which, from the first, such pats and pulls had struck her as the steps and signs of other people's business and even a little as the wriggle or the overflow of their difficulties. What had failed her and what had frightened her on the night of the Exhibition lost themselves at present alike in the impression that any "surprise" now about to burst from Sir Claude would be too big to burst all at once. Any awe that might have sprung from his air of leaving out her stepmother was corrected by the force of a general rule, the odd truth that if Mrs. Beale now never came nor went without making her think of him, it was never, to balance that, the main mark of his own renewed reality to appear to be a reference to Mrs. Beale. To be with Sir Claude was to think of Sir Claude, and that law governed Maisie's mind until, through a sudden lurch of the cab, which had at last taken in Susan and ever so many bundles and almost reached Charing Cross, it popped again somehow into her dizzy head the long-lost image of Mrs. Wix.

It was singular, but from this time she understood and she followed, followed with the sense of an ample filling-out of any void created by symptoms of avoidance and of flight. Her ecstasy was a thing that had yet more of a face than of a back to turn, a pair of eyes still directed to Mrs. Wix even after the slight surprise of their not finding her, as the journey expanded, either at the London station or at the Folkestone hotel. It took few hours to make the child feel that if she was in neither of these places she was at least everywhere else. Maisie had known all along a great deal, but never so much as she was to know from this moment on and as she learned in particular during the couple of days that she was to hang in the air, as it were, over the sea which represented in breezy blueness and with a summer charm a crossing of more spaces than the Channel. It was granted her at this time to arrive at divinations so ample that I shall have no room for the goal if I attempt to trace the stages; as to which therefore I must be content to say that the fullest expression we may give to Sir Claude's conduct is a poor and pale copy of the picture it presented to his young friend. Abruptly, that morning, he had yielded to the action of the idea pumped into him for weeks by Mrs. Wix on lines of approach that she had been capable of the extraordinary art of preserving from entanglement in the fine network of his relations with Mrs. Beale. The breath of her sincerity, blowing without a break, had puffed him up to the flight by which, in the degree I have indicated, Maisie too was carried off her feet. This consisted neither in more nor in less than the brave stroke of his getting off from Mrs. Beale as well as from his wife—of making with the child straight for some such foreign land as would give a support to Mrs. Wix's dream that she might still see his errors renounced and his delinquencies redeemed. It would all be a sacrifice—under eyes that would miss no faintest shade—to what even the strange frequenters of her ladyship's earlier period used to call the real good of the little unfortunate. Maisie's head held a suspicion of much that, during the last long interval, had confusedly, but quite candidly, come and gone in his own; a glimpse, almost awe-stricken in its gratitude, of the miracle her old governess had wrought. That functionary could not in this connexion have been more impressive, even at second-hand, if she had been a prophetess with an open scroll or some ardent abbess speaking with the lips of the Church. She had clung day by day to their plastic associate, plying him with her deep, narrow passion, doing her simple utmost to convert him, and so working on him that he had at last really embraced his fine chance. That the chance was not delusive was sufficiently guaranteed by the completeness with which he could finally figure it out that, in ease of his taking action, neither Ida nor Beale, whose book, on each side, it would only too well suit, would make any sort of row.

It sounds, no doubt, too penetrating, but it was not all as an effect of Sir Claude's betrayals that Maisie was able to piece together the beauty of the special influence through which, for such stretches of time, he had refined upon propriety by keeping, so far as possible, his sentimental interests distinct. She had ever of course in her mind fewer names than conceptions, but it was only with this drawback that she now made out her companion's absences to have had for their ground that he was the lover of her stepmother and that the lover of her stepmother could scarce logically pretend to a superior right to look after her. Maisie had by this time embraced the implication of a kind of natural divergence between lovers and little girls. It was just this indeed that could throw light on the probable contents of the pencilled note deposited on the hall-table in the Regent's Park and which would greet Mrs. Beale on her return. Maisie freely figured it as provisionally jocular in tone, even though to herself on this occasion Sir Claude turned a graver face than he had shown in any crisis but that of putting her into the cab when she had been horrid to him after her parting with the Captain. He might really be embarrassed, but he would be sure, to her view, to have muffled in some bravado of pleasantry the disturbance produced at her father's by the removal of a valued servant. Not that there was n't a great deal too that would n't be in the note—a great deal for which a more comfortable place was Maisie's light little brain, where it hummed away hour after hour and caused the first outlook at Folkestone to swim in a softness of colour and sound. It became clear in this medium that her stepfather had really now only to take into account his entanglement with Mrs. Beale. Was n't he at last disentangled from every one and every thing else? The obstacle to the rupture pressed upon him by Mrs. Wix in the interest of his virtue would be simply that he was in love, or rather, to put it more precisely, that Mrs. Beale had left him no doubt of the degree in which *she* was. She was so much so as to have succeeded in making him accept for a time her infatuated grasp of him and even to some extent the idea of what they yet might do together with a little diplomacy and a good deal of patience. I may not even answer for it that Maisie was not aware of how, in this, Mrs. Beale failed to share his all but insurmountable distaste for their allowing their little charge to breathe the air of their gross irregularity—his contention, in a word, that they should either cease to be irregular or cease to be parental. Their little charge, for herself, had long ago adopted the view that even Mrs. Wix had at one time not thought prohibitively coarse—the view that she was after all, *as* a little charge, morally at home in atmospheres it would be appalling to analyse. If Mrs. Wix, however, ultimately appalled, had now set her heart on strong measures, Maisie, as I have intimated, could also work round both to the reasons for them and to the quite other reasons for that lady's not, as yet at least, appearing in them at first-hand.

Oh decidedly I shall never get you to believe the number of things she saw and the number of secrets she discovered! Why in the world, for instance, could n't Sir Claude have kept it from her—except on the hypothesis of his not caring to—that, when you came to look at it and so far as it was a question of vested interests, he had quite as much right in her as her stepmother, not to say a right that Mrs. Beale was in no position to dispute? He failed at all events of any such successful ambiguity as could keep her, when once they began to look across at France, from regarding even what was least explained as most in the spirit of their old happy times, their rambles and expeditions in the easier better days of their first acquaintance. Never before had she had so the sense of giving him a lead for the sort of treatment of what was between them that would best carry it off, or of his being grateful to her for meeting him so much in the right place. She met him literally at the very point where Mrs. Beale was most to be reckoned with, the point of the jealousy that was sharp in that lady and of the need of their keeping it as long as possible obscure to her that poor Mrs. Wix had still a hand. Yes, she met him too in the truth of the matter that, as her stepmother had had no one else to be jealous of, she had made up for so gross a privation by directing the sentiment to a moral influence. Sir Claude appeared absolutely to convey in a wink that a moral influence capable of pulling a string was after all a moral influence exposed to the scratching out of its eyes; and that, this being the case, there was somebody they could n't afford to leave unprotected before they should see a little better what Mrs. Beale was likely to do. Maisie, true enough, had not to put it into words to rejoin, in the coffee-room, at luncheon: "What *can* she do but come to you if papa does take a step that will amount to legal desertion?" Neither had he then, in answer, to articulate anything but the jollity of their having found a table at a window from which, as they partook of cold beef and apollinaris—for he hinted they would have to save lots of money—they could let their eyes hover tenderly on the far-off white cliffs that so often had signalled to the embarrassed English a promise of safety. Maisie stared at them as if she might really make out after a little a queer dear figure perched on them—a figure as to which she had already the subtle sense that, wherever perched, it would be the very oddest yet seen in France. But it was at least as exciting to feel where Mrs. Wix was n't as it would have been to know where she was, and if she was n't yet at Boulogne this only thickened the plot.

If she was not to be seen that day, however, the evening was marked by an apparition before which, none the less, overstrained suspense folded on the spot its wings. Adjusting her respirations and attaching, under dropped lashes, all her thoughts to a smartness of frock and frill for which she could reflect that she had not appealed in vain to a loyalty in Susan Ash triumphant over the nice things their feverish flight had left behind, Maisie spent on a bench in the garden of the hotel the half-hour before dinner, that mysterious ceremony of the *table d'hôte* for which she had prepared with a punctuality of flutter. Sir Claude, beside her, was occupied with a cigarette and the afternoon papers; and though the hotel was full the garden showed the particular void that ensues upon the sound of the dressing-bell. She had almost had time to weary of the human scene; her own humanity at any rate, in the shape of a smutch on her scanty skirt, had held her so long that as soon as she raised her eyes they rested on a high fair drapery by which smutches were put to shame and which had glided toward her over the grass without her noting its rustle. She followed up its stiff sheen—up and up from the ground, where it had stopped—till at the end of a considerable journey her impression felt the shock of the fixed face which, surmounting it, seemed to offer the climax of the dressed condition. "Why mamma!" she cried the next instant—cried in a tone that, as she sprang to her feet, brought Sir Claude to his own beside her and gave her ladyship, a few yards off, the advantage of their momentary confusion. Poor Maisie's was immense; her mother's drop had the effect of one of the iron shutters that, in evening walks with Susan Ash, she had seen suddenly, at the touch of a spring, rattle down over shining shop-fronts. The light of foreign travel was darkened at a stroke; she had a horrible sense that they were caught; and for the first time of her life in Ida's presence she so far translated an impulse into an invidious act as to clutch straight at the hand of her responsible confederate. It did n't help her that he appeared at first equally hushed with horror; a minute during which, in the empty garden, with its long shadows on the lawn, its blue sea over the hedge and its startled peace in the air, both her elders remained as stiff as tall tumblers filled to the brim and held straight for fear of a spill. At last, in a tone that enriched the whole surprise by its unexpected softness, her mother said to Sir Claude: "Do you mind at all my speaking to her?"

"Oh no; *do* you?" His reply was so long in coming that Maisie was the first to find the right note.

He laughed as he seemed to take it from her, and she felt a sufficient concession in his manner of addressing their visitor. "How in the world did you know we were here?"

His wife, at this, came the rest of the way and sat down on

the bench with a hand laid on her daughter, whom she gracefully drew to her and in whom, at her touch, the fear just
kindled gave a second jump, but now in quite another direction. Sir Claude, on the further side, resumed his seat and
his newspapers, so that the three grouped themselves like
a family party; his connexion, in the oddest way in the world,
almost cynically and in a flash acknowledged, and the mother
patting the child into conformities unspeakable. Maisie
could already feel how little it was Sir Claude and she who
were caught. She had the positive sense of their catching
their relative, catching her in the act of getting rid of her
burden with a finality that showed her as unprecedentedly
relaxed. Oh yes, the fear had dropped, and she had never
been so irrevocably parted with as in the pressure of possession now supremely exerted by Ida's long-gloved and
much-bangled arm. "I went to the Regent's Park"—this was
presently her ladyship's answer to Sir Claude.

"Do you mean to-day?"

"This morning, just after your own call there. That's
how I found you out; that's what has brought me."

Sir Claude considered and Maisie waited. "Whom then
did you see?"

Ida gave a sound of indulgent mockery. "I like your scare.
I know your game. I didn't see the person I risked seeing,
but I had been ready to take my chance of her." She addressed
herself to Maisie; she had encircled her more closely. "I
asked for *you*, my dear, but I saw no one but a dirty parlour-
maid. She was red in the face with the great things that, as
she told me, had just happened in the absence of her mistress; and she luckily had the sense to have made out the
place to which Sir Claude had come to take you. If he had
n't given a false scent I should find you here: that was the
supposition on which I've proceeded." Ida had never been
so explicit about proceeding or supposing, and Maisie,
drinking this in, noted too how Sir Claude shared her fine
impression of it. "I wanted to see you," his wife continued,
"and now you can judge of the trouble I've taken. I had
everything to do in town to-day, but I managed to get off."

Maisie and her companion, for a moment, did justice to this achievement; but Maisie was the first to express it. "I'm glad you wanted to see me, mamma." Then after a concentration more deep and with a plunge more brave: "A little more and you'd have been too late." It stuck in her throat, but she brought it out: "We're going to France."

Ida was magnificent; Ida kissed her on the forehead. "That's just what I thought likely; it made me decide to run down. I fancied that in spite of your scramble you'd wait to cross, and it added to the reason I have for seeing you."

Maisie wondered intensely what the reason could be, but she knew ever so much better than to ask. She was slightly surprised indeed to perceive that Sir Claude did n't, and to hear him immediately enquire: "What in the name of goodness can you have to say to her?"

His tone was not exactly rude, but it was impatient enough to make his wife's response a fresh specimen of the new softness. "That, my dear man, is all my own business."

"Do you mean," Sir Claude asked, "that you wish me to leave you with her?"

"Yes, if you'll be so good; that's the extraordinary request I take the liberty of making." Her ladyship had dropped to a mildness of irony by which, for a moment, poor Maisie was mystified and charmed, puzzled with a glimpse of something that in all the years had at intervals peeped out. Ida smiled at Sir Claude with the strange air she had on such occasions of defying an interlocutor to keep it up as long; her huge eyes, her red lips, the intense marks in her face formed an *éclairage* as distinct and public as a lamp set in a window. The child seemed quite to see in it the very beacon that had lighted her path; she suddenly found herself reflecting that it was no wonder the gentlemen were guided. This must have been the way mamma had first looked at Sir Claude; it brought back the lustre of the time they had outlived. It must have been the way she looked also at Mr. Perriam and Lord Eric; above all it contributed in Maisie's mind to a completer view of that satisfied state of the Captain. Our young lady grasped this idea with a quick lifting of the heart; there was a stillness during which her mother flooded her with a wealth of support to the Captain's striking tribute. This stillness remained long enough unbroken to represent that Sir Claude too might but be gasping again under the spell originally strong for him; so that Maisie quite hoped he would at least say something to show a recognition of how charming she could be.

What he presently said was: "Are you putting up for the night?"

His wife cast grandly about. "Not here—I've come from Dover."

Over Maisie's head, at this, they still faced each other. "You spend the night there?"

"Yes, I brought some things. I went to the hotel and hastily arranged; then I caught the train that whisked me on here. You see what a day I've had of it."

The statement may surprise, but these were really as obliging if not as lucid words as, into her daughter's ears at least, Ida's lips had ever dropped; and there was a quick desire in the daughter that for the hour at any rate they should duly be welcomed as a ground of intercourse. Certainly mamma had a charm which, when turned on, became a large explanation; and the only danger now in an impulse to applaud it would be that of appearing to signalise its rarity. Maisie, however, risked the peril in the geniality of an admission that Ida had indeed had a rush; and she invited Sir Claude to expose himself by agreeing with her that the rush had been even worse than theirs. He appeared to meet this appeal by saying with detachment enough: "You go back there to-night?"

"Oh yes—there are plenty of trains."

Again Sir Claude hesitated; it would have been hard to say if the child, between them, more connected or divided them. Then he brought out quietly: "It will be late for you to knock about. I'll see you over."

"You need n't trouble, thank you. I think you won't deny that I can help myself and that it is n't the first time in my dreadful life that I've somehow managed it." Save for this allusion to her dreadful life they talked there, Maisie noted, as if they were only rather superficial friends; a special effect that she had often wondered at before in the midst of what she supposed to be intimacies. This effect was augmented by the almost casual manner in which her ladyship went on: "I dare say I shall go abroad."

"From Dover do you mean, straight?"

"How straight I can't say. I'm excessively ill."

This for a minute struck Maisie as but a part of the conversation; at the end of which time she became aware that it ought to strike her—though it apparently did n't strike Sir Claude—as a part of something graver. It helped her to twist nearer. "Ill, mamma—really ill?"

She regretted her "really" as soon as she had spoken it; but there could n't be a better proof of her mother's present polish than that Ida showed no gleam of a temper to take it up. She had taken up at other times much tinier things. She only pressed Maisie's head against her bosom and said: "Shockingly, my dear. I must go to that new place."

"What new place?" Sir Claude enquired.

Ida thought, but could n't recall it. "Oh 'Chose,' don't you know?—where every one goes. I want some proper treatment. It's all I've ever asked for on earth. But that's not what I came to say."

Sir Claude, in silence, folded one by one his newspapers; then he rose and stood whacking the palm of his hand with the bundle. "You'll stop and dine with us?"

"Dear no—I can't dine at this sort of hour. I ordered dinner at Dover."

Her ladyship's tone in this one instance showed a certain superiority to those conditions in which her daughter had artlessly found Folkestone a paradise. It was yet not so crushing as to nip in the bud the eagerness with which the latter broke out: "But won't you at least have a cup of tea?"

Ida kissed her again on the brow. "Thanks, love. I had tea before coming." She raised her eyes to Sir Claude. "She *is* sweet!" He made no more answer than if he did n't agree; but Maisie was at ease about that and was still taken up with the joy of this happier pitch of their talk, which put more and more of a meaning into the Captain's version of her ladyship and literally kindled a conjecture that such an admirer might, over there at the other place, be waiting for her to dine. Was the same conjecture in Sir Claude's mind? He partly puzzled her, if it had risen there, by the slight perversity with which he returned to a question that his wife evidently thought she had disposed of.

He whacked his hand again with his paper. "I had really much better take you."

"And leave Maisie here alone?"

Mamma so clearly did n't want it that Maisie leaped at the vision of a Captain who had seen her on from Dover and who, while he waited to take her back, would be hovering just at the same distance at which, in Kensington Gardens, the companion of his walk had herself hovered. Of course, however, instead of breathing any such guess she let Sir Claude reply; all the more that his reply could contribute so much to her own present grandeur. "She won't be alone when she has a maid in attendance."

Maisie had never before had so much of a retinue, and she waited also to enjoy the action of it on her ladyship. "You mean the woman you brought from town?" Ida considered. "The person at the house spoke of her in a way that scarcely made her out company for my child." Her tone was that her child had never wanted, in her hands, for prodigious company. But she as distinctly continued to decline Sir Claude's. "Don't be an old goose," she said charmingly. "Let us alone."

In front of them on the grass he looked graver than Maisie at all now thought the occasion warranted. "I don't see why you can't say it before me."

His wife smoothed one of her daughter's curls. "Say what, dear?"

"Why what you came to say."

At this Maisie at last interposed: she appealed to Sir Claude. "Do let her say it to me."

He looked hard for a moment at his little friend. "How do you know what she may say?"

"She must risk it," Ida remarked.

"I only want to protect you," he continued to the child.

"You want to protect yourself—that's what you mean," his wife replied. "Don't be afraid. I won't touch you."

"She won't touch you—she *won't*!" Maisie declared. She felt by this time that she could really answer for it, and something of the emotion with which she had listened to the Captain came back to her. It made her so happy and so secure that she could positively patronise mamma. She did so in the Captain's very language. "She's good, she's good!" she proclaimed.

"O Lord!"—Sir Claude, at this, let himself go. He appeared to have emitted some sound of derision that was smothered, to Maisie's ears, by her being again embraced by his wife. Ida released her and held her off a little, looking at her with a very queer face. Then the child became aware that their companion had left them and that from the face in question a confirmatory remark had proceeded.

"I *am* good, love," said her ladyship.

XXI

A GOOD deal of the rest of Ida's visit was devoted to explaining, as it were, so extraordinary a statement. This explanation was more copious than any she had yet indulged in, and as the summer twilight gathered and she kept her child in the garden she was conciliatory to a degree that let her need to arrange things a little perceptibly peep out. It was not merely that she explained; she almost conversed; all that was wanting to that was that she should have positively chattered a little less. It was really the occasion of Maisie's life on which her mother was to have most to say to her. That alone was an implication of generosity and virtue, and no great stretch was required to make our young lady feel that she should best meet her and soonest have it over by simply seeming struck with the propriety of her contention. They sat together while the parent's gloved hand sometimes rested sociably on the child's and sometimes gave a corrective pull to a ribbon too meagre or a tress too thick; and Maisie was conscious of the effort to keep out of her eyes the wonder with which they were occasionally moved to blink. Oh there would have been things to blink at if one had let one's self go; and it was lucky they were alone together, without Sir Claude or Mrs. Wix or even Mrs. Beale to catch an imprudent glance. Though profuse and prolonged her ladyship was not exhaustively lucid, and her account of her situation, so far as it could be called descriptive, was a muddle of inconsequent things, bruised fruit of an occasion she had rather too lightly affronted. None of them were really thought out and some were even not wholly insincere. It was as if she had asked outright what better proof could have been wanted of her goodness and her greatness than just this marvellous consent to give up what she had so cherished. It was as if she had said in so many words: "There have been things between us—between Sir Claude and me—which I need n't go into, you little nuisance, because you would n't understand them." It suited her to convey that Maisie had been kept, so far as *she* was concerned or could imagine, in a holy ignorance and that she must take for granted a supreme simplicity. She turned this way and that in the predicament she had sought and from which she could neither retreat with grace nor emerge with credit: she draped herself in the tatters of her impudence, postured to her utmost before the last little triangle of cracked glass to which so many fractures had reduced the polished plate of filial superstition. If neither Sir Claude nor Mrs. Wix was there this was perhaps all the more a pity: the scene had a style of its own that would have qualified it for presentation, especially at such a moment as that of her letting it betray that she quite did think her wretched offspring better placed with Sir Claude than in her own soiled hands. There was at any rate nothing scant either in her admissions or her perversions, the mixture of her fear of what Maisie might undiscoverably think and of the support she at the same time gathered from a necessity of selfishness and a habit of brutality. This habit flushed through the merit she now made, in terms explicit, of not having come to Folkestone to kick up a vulgar row. She had not come to box any ears or to bang any doors or even to use any language: she had come at the worst to lose the thread of her argument in an occasional dumb disgusted twitch of the toggery in which Mrs. Beale's low domestic had had the impudence to serve up Miss Farange. She checked all criticism, not committing herself even so far as for those missing comforts of the schoolroom on which Mrs. Wix had presumed.

"I *am* good—I'm crazily, I'm criminally good. But it won't do for *you* any more, and if I've ceased to contend with him, and with you too, who have made most of the trouble between us, it's for reasons that you'll understand one of these days but too well—one of these days when I hope you'll know what it is to have lost a mother. I'm awfully ill, but you must n't ask me anything about it. If I don't get off somewhere my doctor won't answer for the consequences. He's stupefied at what I've borne—he says it has been put on me because I was formed to suffer. I'm thinking of South Africa, but that's none of your business. You must take your choice—you can't ask me questions if you're so ready to give me up. No, I won't tell you; you can find out for yourself. South Africa's wonderful, they say, and if I do go it must be to give it a fair trial. It must be either one thing or the other; if he takes you, you know, he takes you. I've struck my last blow for you; I can follow you no longer from pillar to post. I must live for myself at last, while there's still a handful left of me. I'm very, very ill; I'm very, very tired; I'm very, very determined. There you have it. Make the most of it. Your frock's too filthy; but I came to sacrifice myself." Maisie looked at the peccant places; there were moments when it was a relief to her to drop her eyes even on anything so sordid. All her interviews, all her ordeals with her mother had, as she had grown older, seemed to have, before any other, the hard quality of duration;

but longer than any, strangely, were these minutes offered to her as so pacific and so agreeably winding up the connexion. It was her anxiety that made them long, her fear of some hitch, some check of the current, one of her ladyship's famous quick jumps. She held her breath; she only wanted, by playing into her visitor's hands, to see the thing through. But her impatience itself made at instants the whole situation swim; there were things Ida said that she perhaps did n't hear, and there were things she heard that Ida perhaps did n't say. "You're all I have, and yet I'm capable of this. Your father wishes you were dead—that, my dear, is what your father wishes. You'll have to get used to it as I've done—I mean to his wishing that *I'm* dead. At all events you see for yourself how wonderful I am to Sir Claude. He wishes me dead quite as much; and I'm sure that if making me scenes about *you* could have killed me—!" It was the mark of Ida's eloquence that she started more hares than she followed, and she gave but a glance in the direction of this one; going on to say that the very proof of her treating her husband like an angel was that he had just stolen off not to be fairly shamed. She spoke as if he had retired on tiptoe, as he might have withdrawn from a place of worship in which he was not fit to be present. "You'll never know what I've been through about you—never, never, never. I spare you everything, as I always have; though I dare say you know things that, if I did (I mean if I knew them)[1] would make me—well, no matter! You're old enough at any rate to know there are a lot of things I don't say that I easily might; though it would do me good, I assure you, to have spoken my mind for once in my life. I don't speak of your father's infamous wife: that may give you a notion of the way I'm letting you off. When I say 'you' I mean your precious friends and

---

[1] The passage in brackets had a confused history. The *Chap Book* serial version reads: '(I mean if I knew you knew them)' and this is followed by the first American edition. The *New Review* serial version reads: '(I mean if you knew them)' and this is followed by the first English edition. The dropping of the two words in these English readings is almost certainly accidental and makes poor sense. The revised text makes adequate sense though the original wording seems preferable.

backers. If you don't do justice to my forbearing, out of delicacy, to mention, just as a last word, about your stepfather, a little fact or two of a kind that really I should only *have* to mention to shine myself in comparison, and after every calumny, like pure gold: if you don't do me *that* justice you'll never do me justice at all!"

Maisie's desire to show what justice she did her had by this time become so intense as to have brought with it an inspiration. The great effect of their encounter had been to confirm her sense of being launched with Sir Claude, to make it rich and full beyond anything she had dreamed, and everything now conspired to suggest that a single soft touch of her small hand would complete the good work and set her ladyship so promptly and majestically afloat as to leave the great seaway clear for the morrow. This was the more the case as her hand had for some moments been rendered free by a marked manœuvre of both of her mother's. One of these capricious members had fumbled with visible impatience in some backward depth of drapery and had presently reappeared with a small article in its grasp. The act had a significance for a little person trained, in that relation, from an early age, to keep an eye on manual motions, and its possible bearing was not darkened by the memory of the handful of gold that Susan Ash would never, never believe Mrs. Beale had sent back—"not she; she's too false and too greedy!"—to the munificent Countess. To have guessed, none the less, that her ladyship's purse might be the real figure of the object extracted from the rustling covert at her rear—this suspicion gave on the spot to the child's eyes a direction carefully distant. It added moreover to the optimism that for an hour could ruffle the surface of her deep diplomacy, ruffle it to the point of making her forget that she had never been safe unless she had also been stupid. She in short forgot her habitual caution in her impulse to adopt her ladyship's practical interests and show her ladyship how perfectly she understood them. She saw without looking that her mother pressed a little clasp; heard, without wanting to, the sharp click that marked the closing portemonnaie from which something had been taken. What this was she just did n't see; it was not too substantial to be locked with ease in the fold of her ladyship's fingers. Nothing was less new to Maisie than the art of not thinking singly, so that at this instant she could both bring out what was on her tongue's end and weigh, as to the object in her mother's palm, the question of its being a sovereign against the question of its being a shilling. No sooner had she begun to speak than she saw that within a few seconds this question would have been settled: she had foolishly checked the rising words of the little speech of presentation to which, under the circumstances, even such a high pride as Ida's had had to give some thought. She had checked it completely—that was the next thing she felt: the note she sounded brought into her companion's eyes a look that quickly enough seemed at variance with presentations.

"That was what the Captain said to me that day, mamma. I think it would have given you pleasure to hear the way he spoke of you."

The pleasure, Maisie could now in consternation reflect, would have been a long time coming if it had come no faster than the response evoked by her allusion to it. Her mother gave her one of the looks that slammed the door in her face; never in a career of unsuccessful experiments had Maisie had to take such a stare. It reminded her of the way that once, at one of the lectures in Glower Street, something in a big jar that, amid an array of strange glasses and bad smells, had been promised as a beautiful yellow was produced as a beautiful black. She had been sorry on that occasion for the lecturer, but she was at this moment sorrier for herself. Oh nothing had ever made for twinges like mamma's manner of saying: "The Captain? What Captain?"

"Why when we met you in the Gardens—the one who took me to sit with him. That was exactly what *he* said."

Ida let it come on so far as to appear for an instant to pick up a lost thread. "What on earth did he say?"

Maisie faltered supremely, but supremely she brought it out. "What you say, mamma—that you're so good."

"What 'I' say?" Ida slowly rose, keeping her eyes on her child, and the hand that had busied itself in her purse conformed at her side and amid the folds of her dress to a certain stiffening of the arm. "I say you're a precious idiot, and I won't have you put words into my mouth!" This was much more peremptory than a mere contradiction. Maisie could only feel on the spot that everything had broken short off and that their communication had abruptly ceased. That was presently proved. "What business have you to speak to me of him?"

Her daughter turned scarlet. "I thought you liked him."

"Him!—the biggest cad in London!" Her ladyship towered again, and in the gathering dusk the whites of her eyes were huge.

Maisie's own, however, could by this time pretty well match them; and she had at least now, with the first flare of anger that had ever yet lighted her face for a foe, the sense of looking up quite as hard as any one could look down. "Well, he was kind about you then; he *was*, and it made me like him. He said things—they were beautiful, they were, they were!" She was almost capable of the violence of forcing this home, for even in the midst of her surge of passion—of which in fact it was a part—there rose in her a fear, a pain, a vision ominous, precocious, of what it might mean for her mother's fate to have forfeited such a loyalty as that. There was literally an instant in which Maisie fully saw—saw madness and desolation, saw ruin and darkness and death. "I've thought of him often since, and I hoped it was with him—with him—!" Here, in her emotion, it failed her, the breath of her filial hope.

But Ida got it out of her. "You hoped, you little horror—?"

"That it was he who's at Dover, that it was he who's to take you. I mean to South Africa," Maisie said with another drop.

Ida's stupefaction, on this, kept her silent unnaturally long, so long that her daughter could not only wonder what was coming, but perfectly measure the decline of every symptom of her liberality. She loomed there in her grandeur, merely dark and dumb; her wrath was clearly still, as it had always been, a thing of resource and variety. What Maisie least expected of it was by this law what now occurred. It melted, in the summer twilight, gradually into pity, and the pity after a little found a cadence to which the renewed click of her purse gave an accent. She had put back what she had taken out. "You're a dreadful dismal deplorable little thing," she murmured. And with this she turned back and rustled away over the lawn.

After she had disappeared Maisie dropped upon the bench again and for some time, in the empty garden and the deeper dusk, sat and stared at the image her flight had still left standing. It had ceased to be her mother only, in the strangest way, that it might become her father, the father of whose wish that she were dead the announcement still lingered in the air. It was a presence with vague edges—it continued to front her, to cover her. But what reality that she need reckon with did it represent if Mr. Farange were, on his side, also going off—going off to America with the Countess, or even only to Spa? That question had, from the house, a sudden gay answer in the great roar of a gong, and at the same moment she saw Sir Claude look out for her from the wide lighted doorway. At this she went to him and he came forward and met her on the lawn. For a minute she was with him there in silence as, just before, at the last, she had been with her mother.

"She's gone?"

"She's gone."

Nothing more, for the instant, passed between them but to move together to the house, where, in the hall, he indulged in one of those sudden pleasantries with which, to the delight of his stepdaughter, his native animation overflowed. "Will Miss Farange do me the honour to accept my arm?"

There was nothing in all her days that Miss Farange had accepted with such bliss, a bright rich element that floated them together to their feast; before they reached which, however, she uttered, in the spirit of a glad young lady taken in to her first dinner a sociable word that made him stop short. "She goes to South Africa."

"To South Africa?" His face, for a moment, seemed to swing for a jump; the next it took its spring into the extreme of hilarity. "Is that what she said?"

"O yes, I did n't *mistake*!" Maisie took to herself *that* credit. "For the climate."

Sir Claude was now looking at a young woman with black hair, a red frock and a tiny terrier tucked under her elbow. She swept past them on her way to the dining-room, leaving an impression of a strong scent which mingled, amid the clatter of the place, with the hot aroma of food. He had become a little graver; he still stopped to talk. "I see—I see." Other people brushed by; he was not too grave to notice them. "Did she say anything else?"

"Oh yes, a lot more."

On this he met her eyes again with some intensity, but only repeating: "I see—I see."

Maisie had still her own vision, which she brought out. "I thought she was going to give me something."

"What kind of a thing?"

"Some money that she took out of her purse and then put back."

Sir Claude's amusement reappeared. "She thought better of it. Dear thrifty soul! How much did she make by that manœuvre?"

Maisie considered. "I did n't see. It was very small."

Sir Claude threw back his head. "Do you mean very little? Sixpence?"

Maisie resented this almost as if, at dinner, she were already bandying jokes with an agreeable neighbour. "It may have been a sovereign."

"Or even," Sir Claude suggested, "a ten-pound note." She flushed at this sudden picture of what she perhaps had lost, and he made it more vivid by adding: "Rolled up in a tight little ball, you know—her way of treating banknotes as if they were curl-papers!" Maisie's flush deepened both with the immense plausibility of this and with a fresh wave of the consciousness that was always there to remind her of his cleverness—the consciousness of how immeasurably

more after all he knew about mamma than she. She had lived
with her so many times without discovering the material of
her curl-papers or assisting at any other of her dealings with
banknotes. The tight little ball had at any rate rolled away
from her for ever—quite like one of the other balls that Ida's
cue used to send flying. Sir Claude gave her his arm again,
and by the time she was seated at table she had perfectly
made up her mind as to the amount of the sum she had
forfeited. Everything about her, however—the crowded
room, the bedizened banquet, the savour of dishes, the
drama of figures—ministered to the joy of life. After dinner
she smoked with her friend—for that was exactly what she
felt she did—on a porch, a kind of terrace, where the red
tips of cigars and the light dresses of ladies made, under the
happy stars, a poetry that was almost intoxicating. They
talked but little, and she was slightly surprised at his asking
for no more news of what her mother had said; but she had
no need of talk—there were a sense and a sound in everything to which words had nothing to add. They smoked and
smoked, and there was a sweetness in her stepfather's silence.
At last he said: "Let us take another turn—but you must go
to bed soon. Oh you know, we're going to have a system!"
Their turn was back into the garden, along the dusky paths
from which they could see the black masts and the red lights
of boats and hear the calls and cries that evidently had to do
with happy foreign travel; and their system was once more to
get on beautifully in this further lounge without a definite exchange. Yet he finally spoke—he broke out as he tossed away
the match from which he had taken a fresh light: "I must go
for a stroll. I'm in a fidget—I must walk it off." She fell in with
this as she fell in with everything; on which he went on: "You
go up to Miss Ash"—it was the name they had started; "you
must see she's not in mischief. Can you find your way alone?"

"Oh yes; I've been up and down seven times." She
positively enjoyed the prospect of an eighth.

Still they did n't separate; they stood smoking together
under the stars. Then at last Sir Claude produced it. "I'm
free—I'm free."

She looked up at him; it was the very spot on which a couple of hours before she had looked up at her mother. "You're free—you're free."

"To-morrow we go to France." He spoke as if he had n't heard her; but it did n't prevent her again concurring.

"To-morrow we go to France."

Again he appeared not to have heard her; and after a moment—it was an effect evidently of the depth of his reflexions and the agitation of his soul—he also spoke as if he had not spoken before. "I'm free—I'm free!"

She repeated her form of assent. "You're free—you're free."

This time he did hear her; he fixed her through the darkness with a grave face. But he said nothing more; he simply stooped a little and drew her to him—simply held her a little and kissed her goodnight; after which, having given her a silent push upstairs to Miss Ash, he turned round again to the black masts and the red lights. Maisie mounted as if France were at the top.

XXII

THE next day it seemed to her indeed at the bottom—down too far, in shuddering plunges, even to leave her a sense, on the Channel boat, of the height at which Sir Claude remained and which had never in every way been so great as when, much in the wet, though in the angle of a screen of canvas, he sociably sat with his stepdaughter's head in his lap and that of Mrs. Beale's housemaid fairly pillowed on his breast. Maisie was surprised to learn as they drew into port that they had had a lovely passage; but this emotion, at Boulogne, was speedily quenched in others, above all in the great ecstasy of a larger impression of life. She was "abroad" and she gave herself up to it, responded to it, in the bright air, before the pink houses, among the bare-legged fishwives and the red-legged soldiers, with the instant certitude of a vocation. Her vocation was to see the world and to thrill with enjoyment of the picture; she had grown older in five minutes and had by the time they reached the hotel recognised in the institutions and manners of France a multitude of affinities and messages. Literally in the course of an hour she found her initiation; a consciousness much quickened by the superior part that, as soon as they had gobbled down a French breakfast—which was indeed a high note in the concert—she observed herself to play to Susan Ash. Sir Claude, who had already bumped against people he knew and who, as he said, had business and letters, sent them out together for a walk, a walk in which the child was avenged, so far as poetic justice required, not only for the loud giggles that in their London trudges used to break from her attendant, but for all the years of her tendency to produce socially that impression of an excess of the queer something which had seemed to waver so widely between innocence and guilt. On the spot, at Boulogne, though there might have been excess there was at least no wavering; she recognised, she understood, she adored and took possession; feeling herself attuned to everything and laying her hand, right and left, on what had simply been waiting for her. She explained to Susan, she laughed at Susan, she towered over Susan; and it was somehow Susan's stupidity, of which she had never yet been so sure, and Susan's bewilderment and ignorance and antagonism, that gave the liveliest rebound to her immediate perceptions and adoptions. The place and the people were all a picture together, a picture that, when they went down to the wide sands, shimmered, in a thousand tints, with the pretty organisation of the *plage*, with the gaiety of spectators and bathers, with that of the language and the weather, and above all with that of our young lady's unprecedented situation. For it appeared to her that no one since the beginning of time could have had such an adventure or, in an hour, so much experience; as a sequel to which she only needed, in order to feel with conscious wonder how the past was changed, to hear Susan, inscrutably aggravated, express a preference for the Edgware Road. The past was so changed and the circle it had formed already so overstepped that on that very afternoon, in the course of another walk, she found herself enquiring of Sir Claude—and without a single scruple—if he were prepared as yet to name the moment at which they should start for Paris. His answer, it must be said, gave her the least little chill.

"Oh Paris, my dear child—I don't quite know about Paris!"

This required to be met, but it was much less to challenge him than for the rich joy of her first discussion of the details of a tour that, after looking at him a minute, she replied: "Well, is n't that the *real* thing, the thing that when one does come abroad—?"

He had turned grave again, and she merely threw that out: it was a way of doing justice to the seriousness of their life. She could n't moreover be so much older since yesterday without reflecting that if by this time she probed a little he would recognise that she had done enough for mere patience. There was in fact something in his eyes that suddenly, to her own, made her discretion shabby. Before she could remedy this he had answered her last question, answered it in the way that, of all ways, she had least expected. "The thing it does n't do not to do? Certainly Paris is charming. But, my dear fellow, Paris eats your head off. I mean it's so beastly expensive."

That note gave her a pang—it suddenly let in a harder light. Were they poor then, that is was *he* poor, really poor beyond the pleasantry of apollinaris and cold beef? They had walked to the end of the long jetty that enclosed the harbour and were looking out at the dangers they had escaped, the grey horizon that was England, the tumbled surface of the sea and the brown smacks that bobbed upon it. Why had he chosen an embarrassed time to make this foreign dash? unless indeed it was just the dash economic, of which she had often heard and on which, after another look at the grey horizon and the bobbing boats, she was ready to turn round with elation. She replied to him quite in his own manner: "I see, I see." She smiled up at him. "Our affairs are involved."

"That's it." He returned her smile. "Mine are not quite so bad as yours; for yours are really, my dear man, in a state I can't see through at all. But mine will do—for a mess."

She thought this over. "But isn't France cheaper than England?"

England, over there in the thickening gloom, looked just then remarkably dear. "I dare say; some parts."

"Then can't we live in those parts?"

There was something that for an instant, in satisfaction of this, he had the air of being about to say and yet not saying. What he presently said was: "This very place is one of them."

"Then we shall live here?"

He didn't treat it quite so definitely as she liked. "Since we've come to save money!"

This made her press him more. "How long shall we stay?"

"Oh three or four days."

It took her breath away. "You can save money in that time?"

He burst out laughing, starting to walk again and taking her under his arm. He confessed to her on the way that she too had put a finger on the weakest of all his weaknesses, the fact, of which he was perfectly aware, that he probably might have lived within his means if he had never done anything for thrift. "It's the happy thoughts that do it," he said; "there's nothing so ruinous as putting in a cheap week." Maisie heard afresh among the pleasant sounds of the closing day that steel click of Ida's change of mind. She thought of the ten-pound note it would have been delightful at this juncture to produce for her companion's encouragement. But the idea was dissipated by his saying irrelevantly, in presence of the next thing they stopped to admire: "We shall stay till she arrives."

She turned upon him. "Mrs. Beale?"

"Mrs. Wix. I've had a wire," he went on. "She has seen your mother."

"Seen mamma?" Maisie stared. "Where in the world?"

"Apparently in London. They've been together."

For an instant this looked ominous—a fear came into her eyes. "Then she has n't gone?"

"Your mother?—to South Africa? I give it up, dear boy," Sir Claude said; and she seemed literally to see him give it up as he stood there and with a kind of absent gaze—absent, that is, from *her* affairs—followed the fine stride and shining limbs of a young fishwife who had just waded out of the sea with her basketful of shrimps. His thought came back to her sooner than his eyes. "But I dare say it's all right. She would n't come if it was n't, poor old thing: she knows rather well what she's about."

This was so reassuring that Maisie, after turning it over, could make it fit into her dream. "Well, what *is* she about?"

He finally stopped looking at the fishwife—he met his companion's enquiry. "Oh you know!" There was something in the way he said it that made, between them, more of an equality than she had yet imagined; but it had also more the effect of raising her up than of letting him down, and what it did with her was shown by the sound of her assent.

"Yes—I know!" What she knew, what she *could* know is by this time no secret to us: it grew and grew at any rate, the rest of that day, in the air of what he took for granted. It was better he should do that than attempt to test her knowledge; but there at the worst was the gist of the matter: it was open between them at last that their great change, as, speaking as if it had already lasted weeks, Maisie called it, was somehow built up round Mrs. Wix. Before she went to bed that night she knew further that Sir Claude, since, as *he* called it, they had been on the rush, had received more telegrams than one. But they separated again without speaking of Mrs. Beale.

Oh what a crossing for the straighteners and the old brown dress—which latter appurtenance the child saw thriftily revived for the possible disasters of travel! The wind got up in the night and from her little room at the inn Maisie could hear the noise of the sea. The next day it was raining and everything different: this was the case even with

Susan Ash, who positively crowed over the bad weather, partly, it seemed, for relish of the time their visitor would have in the boat, and partly to point the moral of the folly of coming to such holes. In the wet, with Sir Claude, Maisie went to the Folkestone packet, on the arrival of which, with many signs of the fray, he made her wait under an umbrella by the quay; whence almost ere the vessel touched, he was to be descried, in quest of their friend, wriggling—that had been his word—through the invalids massed upon the deck. It was long till he reappeared—it was not indeed till every one had landed; when he presented the object of his benevolence in a light that Maisie scarce knew whether to suppose the depth of prostration or the flush of triumph. The lady on his arm, still bent beneath her late ordeal, was muffled in such draperies as had never before offered so much support to so much woe. At the hotel, an hour later, this ambiguity dropped; assisting Mrs. Wix in private to refresh and reinvest herself, Maisie heard from her in detail how little she could have achieved if Sir Claude had n't put it in her power. It was a phrase that in her room she repeated in connexions indescribable: he had put it in her power to have "changes," as she said, of the most intimate order, adapted to climates and occasions so various as to foreshadow in themselves the stages of a vast itinerary. Cheap weeks would of course be in their place after so much money spent on a governess; sums not grudged, however, by this lady's pupil, even on her feeling her own appearance give rise, through the straighteners, to an attention perceptibly mystified. Sir Claude in truth had had less time to devote to it than to Mrs. Wix's; and moreover she would rather be in her own shoes than in her friend's creaking new ones in the event of an encounter with Mrs. Beale. Maisie was too lost in the idea of Mrs. Beale's judgement of so much newness to pass any judgement herself. Besides, after much luncheon and many endearments, the question took quite another turn, to say nothing of the pleasure of the child's quick view that there were other eyes than Susan Ash's to open to what she could show. She could n't show much, alas, till it stopped raining, which it declined to do that day; but this had only the effect of leaving more time for Mrs. Wix's own demonstration. It came as they sat in the little white and gold salon which Maisie thought the loveliest place she had ever seen except perhaps the apartment of the Countess; it came while the hard summer storm lashed the windows and blew in such a chill that Sir Claude, with his hands in his pockets and cigarettes in his teeth, fidgeting, frowning, looking out and turning back, ended by causing a smoky little fire to be made in the dressy little chimney. It came in spite of something that could only be named his air of wishing to put it off; an air that had served him—oh as all his airs served him!—to the extent of his having for a couple of hours confined the conversation to gratuitous jokes and generalities, kept it on the level of the little empty coffee-cups and *petits verres* (Mrs. Wix had two of each!) that struck Maisie, through the fumes of the French fire and the English tobacco, as a token more than ever that they were launched. She felt now, in close quarters and as clearly as if Mrs. Wix had told her, that what this lady had come over for was not merely to be chaffed and to hear her pupil chaffed; not even to hear Sir Claude, who knew French in perfection, imitate the strange sounds emitted by the English folk at the hotel. It was perhaps half an effect of her present renovations, as if her clothes had been somebody's else: she had at any rate never produced such an impression of high colour, of a redness associated in Maisie's mind at *that* pitch either with measles or with "habits." Her heart was not at all in the gossip about Boulogne; and if her complexion was partly the result of the déjeuner and the *petits verres* it was also the brave signal of what she was there to say. Maisie knew when this did come how anxiously it had been awaited by the youngest member of the party. "Her ladyship packed me off—she almost put me into the cab!" That was what Mrs. Wix at last brought out.

XXIII

SIR CLAUDE was stationed at the window; he did n't so much as turn round, and it was left to the youngest of the three to take up the remark. "Do you mean you went to see her yesterday?"

"She came to see *me*. She knocked at my shabby door. She mounted my squalid stair. She told me she had seen you at Folkestone."

Maisie wondered. "She went back that evening?"

"No; yesterday morning. She drove to me straight from the station. It was most remarkable. If I had a job to get off she did nothing to make it worse—she did a great deal to make it better." Mrs. Wix hung fire, though the flame in her face burned brighter; then she became capable of saying: "Her ladyship's kind! She did what I did n't expect."

Maisie, on this, looked straight at her stepfather's back; it might well have been for her at that hour a monument of her ladyship's kindness. It remained, as such, monumentally still, and for a time that permitted the child to ask of their companion: "Did she really help you?"

"Most practically." Again Mrs. Wix paused; again she quite resounded. "She gave me a ten-pound note."

At that, still looking out, Sir Claude, at the window, laughed loud. "So you see, Maisie, we've not quite lost it!"

"Oh no," Maisie responded. "Is n't that too charming?" She smiled at Mrs. Wix. "We know all about it." Then on her friend's showing such blankness as was compatible with such a flush she pursued: "She does want me to have you?"

Mrs. Wix showed a final hesitation, which, however, while Sir Claude drummed on the window-pane, she presently surmounted. It came to Maisie that in spite of his drumming and of his not turning round he was really so much interested as to leave himself in a manner in her hands; which somehow suddenly seemed to her a greater proof than he could have given by interfering. "She wants me to have *you*!" Mrs. Wix declared.

Maisie answered this bang at Sir Claude. "Then that's nice for all of us."

Of course it was, his continued silence sufficiently admitted while Mrs. Wix rose from her chair and, as if to take more of a stand, placed herself, not without majesty, before the fire. The incongruity of her smartness, the circumference of her stiff frock, presented her as really more ready for Paris than any of them. She also gazed hard at Sir Claude's back. "Your wife was different from anything she had ever shown me. She recognises certain proprieties."

"Which? Do you happen to remember?" Sir Claude asked.

Mrs. Wix's reply was prompt. "The importance for Maisie of a gentlewoman, of some one who's not—well, so bad! She objects to a mere maid, and I don't in the least mind telling you what she wants me to do." One thing was clear— Mrs. Wix was now bold enough for anything. "She wants me to persuade you to get rid of the person from Mrs. Beale's."

Maisie waited for Sir Claude to pronounce on this; then she could only understand that he on his side waited, and she felt particularly full of common sense as she met her responsibility. "Oh I don't want Susan with *you*!" she said to Mrs. Wix.

Sir Claude, always from the window, approved. "That's quite simple. I'll take her back."

Mrs. Wix gave a positive jump; Maisie caught her look of alarm. "'Take' her? You don't mean to go over on purpose?"

Sir Claude said nothing for a moment; after which, "Why shouldn't I leave you here?" he enquired.

Maisie, at this, sprang up. "Oh do, oh do, oh do!" The next moment she was interlaced with Mrs. Wix, and the two, on the hearth-rug, their eyes in each other's eyes, considered the plan with intensity. Then Maisie felt the difference of what they saw in it.

"She can surely go back alone: why should you put yourself out?" Mrs. Wix demanded.

"Oh she's an idiot—she's incapable. If anything should happen to her it would be awkward: it was I who brought her—without her asking. If I turn her away I ought with my own hand to place her again exactly where I found her."

Mrs. Wix's face appealed to Maisie on such folly, and her manner, as directed to their companion, had, to her pupil's surprise, an unprecedented firmness. "Dear Sir Claude, I think you're perverse. Pay her fare and give her a sovereign. She has had an experience that she never dreamed of and that will be an advantage to her through life. If she goes wrong on the way it will be simply because she wants to, and, with her expenses and her remuneration—make it even what you like!—you'll have treated her as handsomely as you always treat every one."

This was a new tone—as new as Mrs. Wix's cap; and it could strike a young person with a sharpened sense for latent meanings as the upshot of a relation that had taken on a new character. It brought out for Maisie how much more even than she had guessed her friends were fighting side by side. At the same time it needed so definite a justification that as Sir Claude now at last did face them she at first supposed him merely resentful of excessive familiarity. She was therefore yet more puzzled to see him show his serene beauty untroubled, as well as an equal interest in a matter quite distinct from any freedom but her ladyship's. "Did my wife come alone?" He could ask even that good-humouredly.

"When she called on me?" Mrs. Wix *was* red now: his good humour wouldn't keep down her colour, which for a minute glowed there like her ugly honesty. "No—there was some one in the cab." The only attenuation she could think of was after a minute to add: "But they didn't come up."

Sir Claude broke into a laugh—Maisie herself could guess what it was at: while he now walked about, still laughing, and at the fireplace gave a gay kick to a displaced log, she felt more vague about almost everything than about the drollery of such a "they." She in fact could scarce have told you if it was to deepen or to cover the joke that she bethought herself to observe· "Perhaps it was her maid."

Mrs. Wix gave her a look that at any rate deprecated the wrong tone. "It was not her maid."

"Do you mean there are this time two?" Sir Claude asked as if he had n't heard.

"Two maids?" Maisie went on as if she might assume he had.

The reproach of the straighteners darkened; but Sir Claude cut across it with a sudden: "See here; what do you mean? And what do you suppose *she* meant?"

Mrs. Wix let him for a moment, in silence, understand that the answer to his question, if he did n't take care, might give him more than he wanted. It was as if, with this scruple, she measured and adjusted all she gave him in at last saying: "What she meant was to make me know that you're definitely free. To have that straight from her was a joy I of course had n't hoped for: it made the assurance, and my delight at it, a thing I could really proceed upon. You already know now[1] certainly I'd have started even if she had n't pressed me; you already know what, so long, we've been looking for and what, as soon as she told me of her step taken at Folkestone, I recognised with rapture that we *have*. It's your freedom that makes me right"—she fairly bristled with her logic. "But I don't mind telling you that it's her action that makes me happy!"

"Her action?" Sir Claude echoed. "Why, my dear woman, her action is just a hideous crime. It happens to satisfy our sympathies in a way that's quite delicious; but that does n't in the least alter the fact that it's the most abominable thing ever done. She has chucked our friend here overboard not a bit less than if she had shoved her, shrieking and pleading, out of that window and down two floors to the paving-stones."

Maisie surveyed serenely the parties to the discussion. "Oh your friend here, dear Sir Claude, does n't plead and shriek!"

He looked at her a moment. "Never. Never. That's one, only one, but charming so far as it goes, of about a hundred things we love her for." Then he pursued to Mrs. Wix:

[1] Probably a misprint for 'how'.

"What I can't for the life of me make out is what Ida is *really* up to, what game she was playing in turning to you with that cursed cheek after the beastly way she has used you. Where —to explain her at all—does she fancy she can presently, when we least expect it, take it out of us?"

"She does n't fancy anything, nor want anything out of any one. Her cursed cheek, as you call it, is the best thing I 've ever seen in her. I don't care a fig for the beastly way she used me—I forgive it all a thousand times over!" Mrs. Wix raised her voice as she had never raised it; she quite triumphed in her lucidity. "I understand her, I almost admire her!" she quavered. She spoke as if this might practically suffice; yet in charity to fainter lights she threw out an explanation. "As I 've said, she was different; upon my word I would n't have known her. She had a glimmering, she had an instinct; they brought her. It was a kind of happy thought, and if you could n't have supposed she would ever have had such a thing, why of course I quite agree with you. But she did have it! There!"

Maisie could feel again how a certain rude rightness in this plea might have been found exasperating; but as she had often watched Sir Claude in apprehension of displeasures that did n't come, so now, instead of saying "Oh hell!" as her father used, she observed him only to take refuge in a question that at the worst was abrupt.

"Who *is* it this time, do you know?"

Mrs. Wix tried blind dignity. "Who is what, Sir Claude?"

"The man who stands the cabs. Who was in the one that waited at your door?"

At this challenge she faltered so long that her young friend's pitying conscience gave her a hand. "It was n't the Captain."

This good intention, however, only converted the excellent woman's scruple to a more ambiguous stare; besides of course making Sir Claude go off. Mrs. Wix fairly appealed to him. "Must I really tell you?"

His amusement continued. "Did she make you promise not to?"

Mrs. Wix looked at him still harder. "I mean before Maisie."

Sir Claude laughed again. "Why *she* can't hurt him!"

Maisie felt herself, as it passed, brushed by the light humour of this. "Yes, I can't hurt him."

The straighteners again roofed her over; after which they seemed to crack with the explosion of their wearer's honesty. Amid the flying splinters Mrs. Wix produced a name. "Mr. Tischbein."

There was for an instant a silence that, under Sir Claude's influence and while he and Maisie looked at each other, suddenly pretended to be that of gravity. "We don't know Mr. Tischbein, do we, dear?"

Maisie gave the point all needful thought. "No, I can't place Mr. Tischbein."

It was a passage that worked visibly on their friend. "You must pardon me, Sir Claude," she said with an austerity of which the note was real, "if I thank God to your face that he has in his mercy—I mean his mercy to our charge—allowed me to achieve this act." She gave out a long puff of pain. "It was time!" Then as if still more to point the moral: "I said just now I understood your wife. I said just now I admired her. I stand to it: I did both of those things when I saw how even *she*, poor thing, saw. If you want the dots on the i's you shall have them. What she came to me for, in spite of everything, was that I'm just"—she quavered it out— "well, just clean! What she saw for her daughter was that there must at last be a *decent* person!"

Maisie was quick enough to jump a little at the sound of this implication that such a person was what Sir Claude was not; the next instant, however, she more profoundly guessed against whom the discrimination was made. She was therefore left the more surprised at the complete candour with which he embraced the worst. "If she's bent on decent persons why has she given her to *me*? You don't call me a decent person, and I'll do Ida the justice that *she* never did. I think I'm as indecent as any one and that there's nothing in my behaviour that makes my wife's surrender a bit less ignoble!"

"Don't speak of your behaviour!" Mrs. Wix cried. "Don't say such horrible things; they're false and they're wicked and I forbid you! It's to *keep* you decent that I'm here and that I've done everything I have done. It's to save you—I won't say from yourself, because in yourself you're beautiful and good! It's to save you from the worst person of all. I have n't, after all, come over to be afraid to speak of her! That's the person in whose place her ladyship wants such a person as even me; and if she thought herself, as she as good as told me, not fit for Maisie's company, it's not, as you may well suppose, that she may make room for Mrs. Beale!"

Maisie watched his face as it took this outbreak, and the most she saw in it was that it turned a little white. That indeed made him look, as Susan Ash would have said, queer; and it was perhaps a part of the queerness that he intensely smiled. "You're too hard on Mrs. Beale. She has great merits of her own."

Mrs. Wix, at this, instead of immediately replying, did what Sir Claude had been doing before: she moved across to the window and stared a while into the storm. There was for a minute, to Maisie's sense, a hush that resounded with wind and rain. Sir Claude, in spite of these things, glanced about for his hat; on which Maisie spied it first and, making a dash for it, held it out to him. He took it with a gleam of a "thank-you" in his face, and then something moved her still to hold the other side of the brim; so that, united by their grasp of this object, they stood some seconds looking many things at each other. By this time Mrs. Wix had turned round. "Do you mean to tell me," she demanded, "that you *are* going back?"

"To Mrs. Beale?" Maisie surrendered his hat, and there was something that touched her in the embarrassed, almost humiliated way their companion's challenge made him turn it round and round. She had seen people do that who, she was sure, did nothing else that Sir Claude did. "I can't just say, my dear thing. We'll see about it—we'll talk of it to-morrow. Meantime I must get some air."

Mrs. Wix, with her back to the window, threw up her head to a height that, still for a moment, had the effect of detaining him. "All the air in France, Sir Claude, won't, I think, give you the courage to deny that you're simply afraid of her!"

Oh this time he did look queer; Maisie had no need of Susan's vocabulary to note it! It would have come to her of itself as, with his hand on the door, he turned his eyes from his stepdaughter to her governess and then back again. Resting on Maisie's, though for ever so short a time, there was something they gave up to her and tried to explain. His lips, however, explained nothing; they only surrendered to Mrs. Wix. "Yes. I'm simply afraid of her!" He opened the door and passed out.

It brought back to Maisie his confession of fear of· her mother; it made her stepmother then the second lady about whom he failed of the particular virtue that was supposed most to mark a gentleman. In fact there were three of them, if she counted in Mrs. Wix, before whom he had undeniably quailed. Well, his want of valour was but a deeper appeal to her tenderness. To thrill with response to it she had only to remember all the ladies she herself had, as they called it, funked.

XXIV

IT continued to rain so hard that our young lady's private dream of explaining the Continent to their visitor had to contain a provision for some adequate treatment of the weather. At the *table d'hôte* that evening she threw out a variety of lights: this was the second ceremony of the sort she had sat through, and she would have neglected her privilege and dishonoured her vocabulary—which indeed consisted mainly of the names of dishes—if she had not been proportionately ready to dazzle with interpretations. Preoccupied and overawed, Mrs. Wix was apparently dim: she accepted her pupil's version of the mysteries of the *menu* in a manner that might have struck the child as the depression of a credulity conscious not so much of its needs as of its dimensions. Maisie was soon enough—though it scarce happened before bedtime—confronted again with the different sort of programme for which she reserved her criticism. They remounted together to their sitting-room while Sir Claude, who said he would join them later, remained below to smoke and to converse with the old acquaintances that he met wherever he turned. He had proposed his companions, for coffee, the enjoyment of the *salon de lecture*, but Mrs. Wix had replied promptly and with something of an air that it struck her their own apartments offered them every convenience. They offered the good lady herself, Maisie could immediately observe, not only that of this rather grand reference, which, already emulous, so far as it went, of her pupil, she made as if she had spent her life in salons; but that of a stiff French sofa where she could sit and stare at the faint French lamp, in default of the French clock that had stopped, as for some account of the time Sir Claude would so markedly interpose. Her demeanour accused him so directly of hovering beyond her reach that Maisie sought to divert her by a report of Susan's quaint attitude on the matter of their conversation after lunch. Maisie had mentioned to the young woman for sympathy's sake the plan for her relief, but her disapproval of alien ways appeared, strange to say, only to prompt her to hug her gloom; so that between Mrs. Wix's effect of displacing her and the visible stiffening of her back the child had the sense of a double office and enlarged play for pacific powers.

These powers played to no great purpose, it was true, in keeping before Mrs. Wix the vision of Sir Claude's perversity, which hung there in the pauses of talk and which he himself, after unmistakeable delays, finally made quite lurid by bursting in—it was near ten o'clock—with an object held up in his hand. She knew before he spoke what it was; she knew at least from the underlying sense of all that, since the hour spent after the Exhibition with her father, had not sprung up to reinstate Mr. Farange—she knew it meant

188

a triumph for Mrs. Beale. The mere present sight of Sir Claude's face caused her on the spot to drop straight through her last impression of Mr. Farange a plummet that reached still deeper down than the security of these days of flight. She had wrapped that impression in silence—a silence that had parted with half its veil to cover also, from the hour of Sir Claude's advent, the image of Mr. Farange's wife. But if the object in Sir Claude's hand revealed itself as a letter which he held up very high, so there was something in his mere motion that laid Mrs. Beale again bare. "Here we are!" he cried almost from the door, shaking his trophy at them and looking from one to the other. Then he came straight to Mrs. Wix; he had pulled two papers out of the envelope and glanced at them again to see which was which. He thrust one out open to Mrs. Wix. "Read that." She looked at him hard, as if in fear: it was impossible not to see he was excited. Then she took the letter, but it was not her face that Maisie watched while she read. Neither, for that matter, was it this countenance that Sir Claude scanned: he stood before the fire and, more calmly, now that he had acted, communed in silence with his stepdaughter.

The silence was in truth quickly broken; Mrs. Wix rose to her feet with the violence of the sound she emitted. The letter had dropped from her and lay upon the floor; it had made her turn ghastly white and she was speechless with the effect of it. "It's too abominable—it's too unspeakable!" she then cried.

"Is n't it a charming thing?" Sir Claude asked. "It has just arrived, enclosed in a word of her own. She sends it on to me with the remark that comment's superfluous. I really think it is. That's all you can say."

"She ought n't to pass such a horror about," said Mrs. Wix. "She ought to put it straight in the fire."

"My dear woman, she's not such a fool! It's much too precious." He had picked the letter up and he gave it again a glance of complacency which produced a light in his face. "Such a document"—he considered, then concluded with a slight drop—"such a document is, in fine, a basis!"

"A basis for what?"

"Well—for proceedings."

"Hers?" Mrs. Wix's voice had become outright the voice of derision. "How can *she* proceed?"

Sir Claude turned it over. "How can she get rid of him? Well—she *is* rid of him."

"Not legally." Mrs. Wix had never looked to her pupil so much as if she knew what she was talking about.

"I dare say," Sir Claude laughed; "but she's not a bit less deprived than I!"

"Of the power to get a divorce? It's just your want of the power that makes the scandal of your connexion with her. Therefore it's just her want of it that makes that of hers with you. That's all I contend!" Mrs. Wix concluded with an unparalleled neigh of battle. Oh she did know what she was talking about!

Maisie had meanwhile appealed mutely to Sir Claude, who judged it easier to meet what she did n't say than to meet what Mrs. Wix did.

"It's a letter to Mrs. Beale from your father, my dear, written from Spa and making the rupture between them perfectly irrevocable. It lets her know, and not in pretty language, that, as we technically say, he deserts her. It puts an end for ever to their relations." He ran his eyes over it again, then appeared to make up his mind. "In fact it concerns you, Maisie, so nearly and refers to you so particularly that I really think you ought to see the terms in which this new situation is created for you." And he held out the letter.

Mrs. Wix, at this, pounced upon it; she had grabbed it too soon even for Maisie to become aware of being rather afraid of it. Thrusting it instantly behind her she positively glared at Sir Claude. "See it, wretched man?—the innocent child *see* such a thing? I think you must be mad, and she shall not have a glimpse of it while I'm here to prevent!"

The breadth of her action had made Sir Claude turn red—he even looked a little foolish. "You think it's too bad, eh? But it's precisely because it's bad that it seemed to me it would have a lesson and a virtue for her."

Maisie could do a quick enough justice to his motive to be able clearly to interpose. She fairly smiled at him. "I assure you I can quite believe how bad it is!" She thought of something, kept it back a moment, and then spoke. "I know what's in it!"

He of course burst out laughing and, while Mrs. Wix groaned an "Oh heavens!" replied: "You would n't say that, old boy, if you did! The point I make is," he continued to Mrs. Wix with a blandness now re-established—"the point I make is simply that it sets Mrs. Beale free."

She hung fire but an instant. "Free to live with *you*?"

"Free not to live, not to pretend to live, with her husband."

"Ah they're mighty different things!"—a truth as to which her earnestness could now with a fine inconsequent look invite the participation of the child.

Before Maisie could commit herself, however, the ground was occupied by Sir Claude, who, as he stood before their visitor with an expression half rueful, half persuasive, rubbed his hand sharply up and down the back of his head. "Then why the deuce do you grant so—do you, I may even say, rejoice so—that by the desertion of my own precious partner I'm free?"

Mrs. Wix met this challenge first with silence, then with a demonstration the most extraordinary, the most unexpected. Maisie could scarcely believe her eyes as she saw the good lady, with whom she had associated no faintest shade of any art of provocation, actually, after an upward grimace, give Sir Claude a great giggling insinuating naughty slap. "You wretch—you *know* why!" And she turned away. The face that with this movement she left him to present to Maisie was to abide with his stepdaughter as the very image of stupefaction; but the pair lacked time to communicate either amusement or alarm before their admonisher was upon them again. She had begun in fact to show infinite variety and she flashed about with a still quicker change of tone. "Have you brought me that thing as a pretext for your going over?"

Sir Claude braced himself. "I can't, after such news, in common decency not go over. I mean, don't you know, in common courtesy and humanity. My dear lady, you can't chuck a woman that way, especially taking the moment when she has been most insulted and wronged. A fellow must behave like a gentleman, damn it, dear good Mrs. Wix. We did n't come away, we two, to hang right on, you know: it was only to try our paces and just put in a few days that might prove to every one concerned that we're in earnest. It's exactly because we're in earnest that, dash it, we need n't be so awfully particular. I mean, don't you know, we need n't be so awfully afraid." He showed a vivacity, an intensity of argument, and if Maisie counted his words she was all the more ready to swallow after a single swift gasp those that, the next thing, she became conscious he paused for a reply to. "We did n't come, old girl, did we," he pleaded straight, "to stop right away for ever and put it all in *now*?"

Maisie had never doubted she could be heroic for him. "Oh no!" It was as if she had been shocked at the bare thought. "We're just taking it as we find it." She had a sudden inspiration, which she backed up with a smile. "We're just seeing what we can afford." She had never yet in her life made any claim for herself, but she hoped that this time, frankly, what she was doing would somehow be counted to her. Indeed she felt Sir Claude *was* counting it, though she was afraid to look at him—afraid she should show him tears. She looked at Mrs. Wix; she reached her maximum. "I don't think I ought to be bad to Mrs. Beale."

She heard, on this, a deep sound, something inarticulate and sweet, from Sir Claude; but tears were what Mrs. Wix did n't scruple to show. "Do you think you ought to be bad to *me*?" The question was the more disconcerting that Mrs. Wix's emotion did n't deprive her of the advantage of her effect. "If you see that woman again you're lost!" she declared to their companion.

Sir Claude looked at the moony globe of the lamp; he seemed to see for an instant what seeing Mrs. Beale would consist of. It was also apparently from this vision that he drew strength to return: "Her situation, by what has happened, is completely changed; and it's no use your trying to prove to me that I need n't take any account of that."

"If you see that woman you're lost!" Mrs. Wix with greater force repeated.

"Do you think she'll not let me come back to you? My dear lady, I leave you here, you and Maisie, as a hostage to fortune, and I promise you by all that's sacred that I shall be with you again at the very latest on Saturday. I provide you with funds; I install you in these lovely rooms; I arrange with the people here that you be treated with every attention and supplied with every luxury. The weather, after this, will mend; it will be sure to be exquisite. You'll both be as free as air and you can roam all over the place and have tremendous larks. You shall have a carriage to drive you; the whole house shall be at your call. You'll have a magnificent position." He paused, he looked from one of his companions to the other as to see the impression he had made. Whether or no he judged it adequate he subjoined after a moment: "And you'll oblige me above all by not making a fuss."

Maisie could only answer for the impression on herself, though indeed from the heart even of Mrs. Wix's rigour there floated to her sense a faint fragrance of depraved concession. Maisie had her dumb word for the show such a speech could make, for the irresistible charm it could take from his dazzling sincerity; and before she could do anything but blink at excess of light she heard this very word sound on Mrs. Wix's lips, just as if the poor lady had guessed it and wished, snatching it from her, to blight it like a crumpled flower. "You're dreadful, you're terrible, for you know but too well that it's not a small thing to me that you should address me in terms that are princely!" Princely was what he stood there and looked and sounded; that was what Maisie for the occasion found herself reduced to simple worship of him for being. Yet strange to say too, as Mrs. Wix went on, an echo rang within her that matched the echo she had herself just produced. "How much you must *want* to see her to say such things as that and to be ready to do so much for the poor little likes of Maisie and me! She has a hold on you, and you know it, and you want to feel it again and—God knows, or at least *I* do, what's your motive and desire—enjoy it once more and give yourself up to it! It does n't matter if it's one day or three: enough is as good as a feast and the lovely time you'll have with her is something you're willing to pay for! I dare say you'd like me to believe that your pay is to get her to give you up; but that's a matter on which I strongly urge you not to put down your money in advance. Give *her* up first. Then pay her what you please!"

Sir Claude took this to the end, though there were things in it that made him colour, called into his face more of the apprehension than Maisie had ever perceived there of a particular sort of shock. She had an odd sense that it was the first time she had seen any one but Mrs. Wix really and truly scandalised, and this fed her inference, which grew and grew from moment to moment, that Mrs. Wix was proving more of a force to reckon with than either of them had allowed so much room for. It was true that, long before, she had obtained a "hold" of him, as she called it, different in kind from that obtained by Mrs. Beale and originally by her ladyship. But Maisie could quite feel with him now that he had really not expected this advantage to be driven so home. Oh they had n't at all got to where Mrs. Wix would stop, for the next minute she was driving harder than ever. It was the result of his saying with a certain dryness, though so kindly that what most affected Maisie in it was his patience: "My dear friend, it's simply a matter in which I must judge for myself. You've judged *for* me, I know, a good deal, of late, in a way that I appreciate, I assure you, down to the ground. But you can't do it always; no one can do that for another, don't you see, in every case. There are exceptions, particular cases that turn up and that are awfully delicate. It would be too easy if I could shift it all off on you: it would be allowing you to incur an amount of responsibility that I should simply become quite ashamed of. You'll find, I'm sure, that you'll have quite as much as you'll enjoy if you'll be so good as to accept the situation as circumstances happen to make it for you and to stay here with our friend, till I rejoin you, on the footing of as much pleasantness and as much comfort—and I think I have a right to add, to both of you, of as much faith in *me*—as possible."

Oh he was princely indeed: that came out more and more with every word he said and with the particular way he said it, and Maisie could feel his monitress stiffen almost with anguish against the increase of his spell and then hurl herself as a desperate defence from it into the quite confessed poorness of violence, of iteration. "You're afraid of her—afraid, afraid, afraid! Oh dear, oh dear, oh dear!" Mrs. Wix wailed it with a high quaver, then broke down into a long shudder of helplessness and woe. The next minute she had flung herself again on the lean sofa and had burst into a passion of tears.

Sir Claude stood and looked at her a moment; he shook his head slowly, altogether tenderly. "I've already admitted it—I'm in mortal terror; so we'll let that settle the question. I think you had best go to bed," he added; "you've had a tremendous day and you must both be tired to death. I shall not expect you to concern yourselves in the morning with my movements. There's an early boat on; I shall have cleared out before you're up; and I shall moreover have dealt directly and most effectively, I assure you, with the haughty but not quite hopeless Miss Ash." He turned to his stepdaughter as if at once to take leave of her and give her a sign of how, through all tension and friction, they were still united in such a way that she at least need n't worry. "Maisie boy!"—he opened his arms to her. With her culpable lightness she flew into them and, while he kissed her, chose the soft method of silence to satisfy him, the silence that after battles of talk was the best balm she could offer his wounds. They held each other long enough to reaffirm intensely their vows; after which they were almost forced apart by Mrs. Wix's jumping to her feet.

Her jump, either with a quick return or with a final lapse of courage, was also to supplication almost abject. "I beseech you not to take a step so miserable and so fatal. I know her but too well, even if you jeer at me for saying it; little as I've seen her I know her, I know her. I know what she'll do—I see it as I stand here. Since you're afraid of her it's the mercy of heaven. Don't, for God's sake, be afraid to show it, to profit by it and to arrive at the very safety that it gives you. *I'm* not afraid of her, I assure you; you must already have seen for yourself that there's nothing I'm afraid of now. Let me go to her—*I'll* settle her and I'll take that woman back without a hair of her touched. Let me put in the two or three days—let me wind up the connexion. You stay here with Maisie, with the carriage and the larks and the luxury; then I'll return to you and we'll go off together—we'll live together without a cloud. Take me, take me," she went on and on—the tide of her eloquence was high. "Here I am; I know what I am and what I ain't; but I say boldly to the face of you both that I'll do better for you, far, than ever she'll even try to. I say it to yours, Sir Claude, even though I owe you the very dress on my back and the very shoes on my feet. I owe you everything—that's just the reason; and to pay it back, in profusion, what can that be but what I want? Here I am, here I am!"—she spread herself into an exhibition that, combined with her intensity and her decorations, appeared to suggest her for strange offices and devotions, for ridiculous replacements and substitutions. She manipulated her gown as she talked, she insisted on the items of her debt. "I have nothing of my own, I know —no money, no clothes, no appearance, no anything, nothing but my hold of this little one truth, which is all in the world I can bribe you with: that the pair of you are more to me than all besides, and that if you'll let me help you and save you, make what you both want possible in the one way it *can* be, why, I'll work myself to the bone in your service!"

Sir Claude wavered there without an answer to this magnificent appeal; he plainly cast about for one, and in no small agitation and pain. He addressed himself in his quest, however, only to vague quarters until he met again, as he so frequently and actively met it, the more than filial gaze of his intelligent little charge. That gave him—poor plastic and dependent male—his issue. If she was still a child she was yet of the sex that could help him out. He signified as much by a renewed invitation to an embrace. She freshly sprang to him and again they inaudibly conversed. "Be nice to her, be nice to her," he at last distinctly articulated; "be nice to her as you've not even been to *me*!" On which, without another look at Mrs. Wix, he somehow got out of the room, leaving Maisie under the slight oppression of these words as well as of the idea that he had unmistakeably once more dodged.

XXV

EVERY single thing he had prophesied came so true that it was after all no more than fair to expect quite as much for what he had as good as promised. His pledges they could verify to the letter, down to his very guarantee that a way would be found with Miss Ash. Roused in the summer dawn and vehemently squeezed by that interesting exile, Maisie fell back upon her couch with a renewed appreciation of his policy, a memento of which, when she rose later on to dress, glittered at her from the carpet in the shape of a sixpence that had overflowed from Susan's pride of possession. Sixpences really, for the forty-eight hours that followed, seemed to abound in her life; she fancifully computed the number of them represented by such a period of "larks." The number was not kept down, she presently noticed, by any scheme of revenge for Sir Claude's flight which should take on Mrs. Wix's part the form of a refusal to avail herself of the facilities he had so bravely ordered. It was in fact impossible to escape them; it was in the good lady's own phrase ridiculous to go on foot when you had a carriage prancing at the door. Everything about them pranced: the very waiters even as they presented the dishes to which, from a similar sense of the absurdity oi perversity, Mrs. Wix helped herself with a freedom that spoke to Maisie quite as much of her depletion as of her logic. Her appetite was a sign to her companion of a great many things and testified no less on the whole to her general than to her particular condition. She had arrears of dinner to make up, and it was touching that in a dinnerless state her moral passion should have burned so clear. She partook largely as a refuge from depression, and yet the opportunity to partake was just a mark of the sinister symptoms that depressed her. The affair was in short a combat, in which the baser element triumphed, between her refusal to be bought off and her consent to be clothed and fed. It was not at any rate to be gainsaid that there was comfort for her in the developments of France; comfort so great as to leave Maisie free to take with her all the security for granted and brush all the danger aside. That was the way to carry out in detail Sir Claude's injunction to be "nice"; that was the way, as well, to look, with her, in a survey of the pleasures of life abroad, straight over the head of any doubt.

They shrank at last, all doubts, as the weather cleared up: it had an immense effect on them and became quite as lovely as Sir Claude had engaged. This seemed to have put him so into the secret of things, and the joy of the world so waylaid the steps of his friends, that little by little the spirit of hope filled the air and finally took possession of the scene. To drive on the long cliff was splendid, but it was perhaps better still to creep in the shade—for the sun was strong—along the many-coloured and many-odoured *port* and through the streets in which, to English eyes, everything that was the same was a mystery and everything that was different a joke. Best of all was to continue the creep up the long Grand' Rue to the gate of the *haute ville* and, passing beneath it, mount to the quaint and crooked rampart, with its rows of trees, its quiet corners and friendly benches where brown old women in such white-frilled caps and such long gold earrings sat and knitted or snoozed, its little yellow-faced houses that looked like the homes of misers or of priests and its dark château where small soldiers lounged on the bridge that stretched across an empty moat and military washing hung from the windows of towers. This was a part of the place that could lead Maisie to enquire if it did n't just meet one's idea of the middle ages; and since it was rather a satisfaction than a shock to perceive, and not for the first time, the limits in Mrs. Wix's mind of the historic imagination, that only added one more to the variety of kinds of insight that she felt it her own present mission to show. They sat together on the old grey bastion; they looked down on the little new town which seemed to them quite as old, and across at the great dome and the high gilt Virgin of the church that, as they gathered, was famous and that pleased them by its unlikeness to any place in which they had worshipped. They wandered in this temple afterwards and Mrs. Wix confessed that for herself she had probably made a fatal mistake early in life in not being a Catholic. Her confession in its turn caused Maisie to wonder rather interestedly what degree of lateness it was that shut the door against an escape from such an error. They went back to the rampart on the second morning— the spot on which they appeared to have come furthest in the journey that was to separate them from everything objection- able in the past: it gave them afresh the impression that had most to do with their having worked round to a confidence that on Maisie's part was determined and that she could see to be on her companion's desperate. She had had for many hours the sense of showing Mrs. Wix so much that she was comparatively slow to become conscious of being at the same time the subject of a like aim. The business went the faster, however, from the moment she got her glimpse of it; it then fell into its place in her general, her habitual view of the particular phenomenon that, had she felt the need of words for it, she might have called her personal relation to her knowledge. This relation had never been so lively as during the time she waited with her old governess for Sir Claude's reappearance, and what made it so was exactly that Mrs. Wix struck her as having a new suspicion of it. Mrs. Wix had never yet had a suspicion—this was certain—so calculated to throw her pupil, in spite of the closer union of such adventurous hours, upon the deep defensive. Her pupil made out indeed as many marvels as she had made out on the rush to Folkestone; and if in Sir Claude's company on that occasion Mrs. Wix was the constant implication, so in Mrs. Wix's, during these hours, Sir Claude was—and most of all through long pauses—the perpetual, the insurmountable theme. It all took them back to the first flush of his marriage and to the place he held in the schoolroom at that crisis of love and pain; only he had himself blown to a much bigger balloon the large consciousness he then filled out.

They went through it all again, and indeed while the interval dragged by the very weight of its charm they went, in spite of defences and suspicions, through everything. Their intensified clutch of the future throbbed like a clock ticking seconds; but this was a timepiece that inevitably, as well, at the best, rang occasionally a portentous hour. Oh there were several of these, and two or three of the worst on the old city-wall where everything else so made for peace. There was nothing in the world Maisie more wanted than to be as nice to Mrs. Wix as Sir Claude had desired; but it was exactly because this fell in with her inveterate instinct of keeping the peace that the instinct itself was quickened. From the moment it was quickened, however, it found other work, and that was how, to begin with, she produced the very complication she most sought to avert. What she had essentially done, these days, had been to read the unspoken into the spoken; so that thus, with accumulations, it had become more definite to her that the unspoken was, unspeakably, the completeness of the sacrifice of Mrs. Beale. There were times when every minute that Sir Claude stayed away was like a nail in Mrs. Beale's coffin. That brought back to Maisie—it was a roundabout way—the beauty and antiquity of her connexion with the flower of the Overmores as well as that lady's own grace and charm, her peculiar prettiness and cleverness and even her peculiar tribulations. A hundred things hummed at the back of her head, but two of these were simple enough. Mrs. Beale was by the way, after all, just her stepmother and her relative. She was just—and partly for that very reason—Sir Claude's greatest intimate ("lady-intimate" was Maisie's term) so that what together they were on Mrs. Wix's prescription to give up and break short off with was for one of them his particular favourite and for the other her father's wife. Strangely, indescribably her perception of reasons kept pace with her sense of trouble; but there was something in her that, without a supreme effort not to be shabby, could n't take the reasons for granted. What it comes to perhaps for ourselves is that, disinherited and denuded as we have seen her, there still lingered in her life an echo of parental influence—she was still reminiscent of one of the sacred lessons of home. It was the only one she retained, but luckily she retained it with force. She enjoyed in a word an ineffaceable view of the fact that there were things papa called mamma and mamma called papa a low sneak for doing or for not doing. Now this rich memory gave her a name that she dreaded to invite to the lips of Mrs. Beale: she should personally wince so just to hear it. The very sweetness of the foreign life she was steeped in added with each hour of Sir Claude's absence to the possibility of such pangs. She watched beside Mrs. Wix the great golden Madonna, and one of the ear-ringed old women who had been sitting at the end of their bench got up and pottered away.

"Adieu mesdames!" said the old woman in a little cracked civil voice—a demonstration by which our friends were so affected that they bobbed up and almost curtseyed to her. They subsided again, and it was shortly after, in a summer hum of French insects and a phase of almost somnolent reverie, that Maisie most had the vision of what it was to shut out from such a perspective so appealing a participant. It had not yet appeared so vast as at that moment, this prospect of statues shining in the blue and of courtesy in romantic forms.

"Why after all should we have to choose between you? Why should n't we be four?" she finally demanded.

Mrs. Wix gave the jerk of a sleeper awakened or the start even of one who hears a bullet whiz at the flag of truce. Her stupefaction at such a breach of the peace delayed for a moment her answer. "Four improprieties, do you mean?

Because two of us happen to be decent people! Do I gather you to wish that I should stay on with you even if that woman *is* capable—?"

Maisie took her up before she could further phrase Mrs. Beale's capability. "Stay on as *my* companion—yes. Stay on as just what you were at mamma's. Mrs. Beale *would* let you!" the child said.

Mrs. Wix had by this time fairly sprung to her arms. "And who, I'd like to know, would let Mrs. Beale? Do you mean, little unfortunate, that *you* would?"

"Why not, if now she's free?"

"Free? Are you imitating *him*? Well, if Sir Claude's old enough to know better, upon my word I think it's right to treat you as if you also were. You'll have to, at any rate—to know better—if that's the line you're proposing to take." Mrs. Wix had never been so harsh; but on the other hand Maisie could guess that she herself had never appeared so wanton. What was underlying, however, rather overawed than angered her; she felt she could still insist—not for contradiction, but for ultimate calm. Her wantonness meanwhile continued to work upon her friend, who caught again, on the rebound, the sound of deepest provocation. "Free, free, free? If she's as free as *you* are, my dear, she's free enough, to be sure!"

"As I am?"—Maisie, after reflexion and despite whatever of portentous this seemed to convey, risked a critical echo.

"Well," said Mrs. Wix, "nobody, you know, is free to commit a crime."

"A crime!" The word had come out in a way that made the child sound it again.

"You'd commit as great a one as their own—and so should I—if we were to condone their immorality by our presence."

Maisie waited a little; this seemed so fiercely conclusive. "Why is it immorality?" she nevertheless presently enquired.

Her companion now turned upon her with a reproach softer because it was somehow deeper. "You're too unspeakable! Do you know what we're talking about?"

In the interest of ultimate calm Maisie felt that she must

be above all clear. "Certainly; about their taking advantage of their freedom."

"Well, to do what?"

"Why, to live with us."

Mrs. Wix's laugh, at this, was literally wild. "'Us?' Thank you!"

"Then to live with *me*."

The words made her friend jump. "You give me up? You break with me for ever? You turn me into the street?"

Maisie, though gasping a little, bore up under the rain of challenges. "Those, it seems to me, are the things you do to *me*."

Mrs. Wix made little of her valour. "I can promise you that, whatever I do, I shall never let you out of my sight! You ask me why it's immorality when you've seen with your own eyes that Sir Claude has felt it to be so to that dire extent that, rather than make you face the shame of it, he has for months kept away from you altogether? Is it any more difficult to see that the first time he tries to do his duty he washes his hands of *her*—takes you straight away from her?"

Maisie turned this over, but more for apparent consideration than from any impulse to yield too easily. "Yes, I see what you mean. But at that time they were n't free." She felt Mrs. Wix rear up again at the offensive word, but she succeeded in touching her with a remonstrant hand. "I don't think you know how free they've become."

"I know, I believe, at least as much as you do!"

Maisie felt a delicacy but overcame it. "About the Countess?"

"Your father's—temptress?" Mrs. Wix gave her a side-long squint. "Perfectly. She pays him!"

"Oh *does* she?" At this the child's countenance fell: it seemed to give a reason for papa's behaviour and place it in a more favourable light. She wished to be just. "I don't say she's not generous. She was so to me."

"How, to you?"

"She gave me a lot of money."

Mrs. Wix stared. "And pray what did you do with a lot of money?"

"I gave it to Mrs. Beale."

"And what did Mrs. Beale do with it?"

"She sent it back."

"To the Countess? Gammon!" said Mrs. Wix. She disposed of that plea as effectually as Susan Ash.

"Well, I don't care!" Maisie replied. "What I mean is that you don't know about the rest."

"The rest? What rest?"

Maisie wondered how she could best put it. "Papa kept me there an hour."

"I do know—Sir Claude told me. Mrs. Beale had told him."

Maisie looked incredulity. "How could she—when I did n't speak of it?"

Mrs. Wix was mystified. "Speak of what?"

"Why, of her being so frightful."

"The Countess? Of course she's frightful!" Mrs. Wix returned. After a moment she added· "That's why she pays him."

Maisie pondered. "It's the best thing about her then—if she gives him as much as she gave *me*."

"Well, it's not the best thing about *him*! Or rather perhaps it *is* too!" Mrs. Wix subjoined.

"But she's awful—really and truly,' Maisie went on.

Mrs. Wix arrested her. "You need n't go into details!" It was visibly at variance with this injunction that she yet enquired: "How does that make it any better?"

"Their living with me? Why for the Countess—and for her whiskers!—he has put me off on them. I understood him," Maisie profoundly said.

"I hope then he understood you. It's more than I do!" Mrs. Wix admitted.

This was a real challenge to be plainer, and our young lady immediately became so. "I mean it is n't a crime."

"Why then did Sir Claude steal you away?"

"He did n't steal—he only borrowed me. I knew it was n't for long," Maisie audaciously professed.

"You must allow me to reply to that," cried Mrs. Wix, "that you knew nothing of the sort, and that you rather b asely failed to back me up last night when you pretended so plump that you did! You hoped in fact, exactly as much as I did and as in my senseless passion I even hope now, that this may be the beginning of better things."

Oh yes, Mrs. Wix was indeed, for the first time, sharp; so that there at last stirred in our heroine the sense not so much of being proved disingenuous as of being precisely accused of the meanness that had brought everything down on her through her very desire to shake herself clear of it. She suddenly felt herself swell with a passion of protest. "I never, *never* hoped I was n't going again to see Mrs. Beale! I did n't, I did n't, I did n't!" she repeated. Mrs. Wix bounced about with a force of rejoinder of which she also felt that she must anticipate the concussion and which, though the good lady was evidently charged to the brim, hung fire long enough to give time for an aggravation. "She's beautiful and I love her! I love her and she's beautiful!"

"And I'm hideous and you hate *me*?" Mrs. Wix fixed her a moment, then caught herself up. "I won't embitter you by absolutely accusing you of that; though, as for my being hideous, it's hardly the first time I've been told so! I know it so well that even if I have n't whiskers—have I?—I dare say there are other ways in which the Countess is a Venus to me! My pretensions must therefore seem to you monstrous —which comes to the same thing as your not liking me. But do you mean to go so far as to tell me that you *want* to live with them in their sin?"

"You know what I want, you know what I want!"— Maisie spoke with the shudder of rising tears.

"Yes, I do; you want me to be as bad as yourself! Well, I won't. There! Mrs. Beale's as bad as your father!" Mrs. Wix went on.

"She's not!—she's not!" her pupil almost shrieked in retort.

"You mean because Sir Claude at least has beauty and wit and grace? But he pays just as the Countess pays!" Mrs. Wix, who now rose as she spoke, fairly revealed a latent cynicism.

It raised Maisie also to her feet; her companion had walked off a few steps and paused. The two looked at each other as they had never looked, and Mrs. Wix seemed to flaunt there in her finery. "Then does n't he pay *you* too?" her unhappy charge demanded.

At this she bounded in her place. "Oh you incredible little waif!" She brought it out with a wail of violence; after which, with another convulsion, she marched straight away.

Maisie dropped back on the bench and burst into sobs.

XXVI

NOTHING so dreadful of course could be final or even for many minutes prolonged: they rushed together again too soon for either to feel that either had kept it up, and though they went home in silence it was with a vivid perception for Maisie that her companion's hand had closed upon her. That hand had shown altogether, these twenty-four hours, a new capacity for closing, and one of the truths the child could least resist was that a certain greatness had now come to Mrs. Wix. The case was indeed that the quality of her motive surpassed the sharpness of her angles; both the combination and the singularity of which things, when in the afternoon they used the carriage, Maisie could borrow from the contemplative hush of their grandeur the freedom to feel to the utmost. She still bore the mark of the tone in which her friend had thrown out that threat of never losing sight of her. This friend had been converted in short from feebleness to force; and it was the light of her new authority that showed from how far she had come. The threat in question, sharply exultant, might have produced defiance; but before anything so ugly could happen another process had insidiously forestalled it. The moment at which this process had begun to mature was that of Mrs. Wix's breaking out with a dignity attuned to their own apartments and with an advantage now measurably gained. They had ordered coffee after luncheon, in the spirit of Sir Claude's provision, and it was served to them while they awaited their equipage in the white and gold saloon. It was flanked moreover with a couple of liqueurs, and Maisie felt that Sir Claude could scarce have been taken more at his word had it been followed by anecdotes and cigarettes. The influence of these luxuries was at any rate in the air. It seemed to her while she tiptoed at the chimney-glass, pulling on her gloves and with a motion of her head shaking a feather into place, to have had something to do with Mrs. Wix's suddenly saying: "Have n't you really and truly *any* moral sense?"

Maisie was aware that her answer, though it brought her down to her heels, was vague even to imbecility, and that this was the first time she had appeared to practise with Mrs. Wix an intellectual inaptitude to meet her—the infirmity to which she had owed so much success with papa and mamma. The appearance did her injustice, for it was not less through her candour than through her playfellow's pressure than after this the idea of a moral sense mainly coloured their intercourse. She began, the poor child, with scarcely knowing what it was; but it proved something that, with scarce an outward sign save her surrender to the swing of the carriage, she could, before they came back from their drive, strike up a sort of acquaintance with. The beauty of the day only deepened, and the splendour of the afternoon sea, and the haze of the far headlands, and the taste of the sweet air. It was the coachman indeed who, smiling and cracking his whip, turning in his place, pointing to invisible objects and uttering unintelliglble sounds—all, our tourists recognised, strict features of a social order principally devoted to language: it was this polite person, I say, who made their excursion fall so much short that their return left them still a stretch of the long daylight and an hour that, at his obliging suggestion, they spent on foot by the shining sands. Maisie had seen the *plage* the day before with Sir Claude, but that was a reason the more for showing on the spot to Mrs. Wix that it was, as she said, another of the places on her list and of the things of which she knew the French name. The bathers, so late, were absent and the tide was low; the sea-pools twinkled in the sunset and there were dry places as well, where they could sit again and admire and expatiate: a circumstance that, while they listened to the lap of the waves, gave Mrs. Wix a fresh support for her challenge. "Have you absolutely none at all?"

She had no need now, as to the question itself at least, to be specific; that on the other hand was the eventual result of their quiet conjoined apprehension of the thing that— well, yes, since they must face it—Maisie absolutely and appallingly had so little of. This marked more particularly the moment of the child's perceiving that her friend had risen to a level which might—till superseded at all events— pass almost for sublime. Nothing more remarkable had taken place in the first heat of her own departure, no act of perception less to be overtraced by our rough method, than her vision, the rest of that Boulogne day, of the manner in which she figured. I so despair of courting her noiseless mental footsteps here that I must crudely give you my word for its being from this time forward a picture literally present to her. Mrs. Wix saw her as a little person knowing so extraordinarily much that, for the account to be taken of it, what she still did n't know would be ridiculous if it had n't been embarrassing. Mrs. Wix was in truth more than ever qualified to meet embarrassment; I am not sure that Maisie had not even a dim discernment of the queer law of her own life that made her educate to that sort of proficiency those elders with whom she was concerned. She promoted, as it were, their development; nothing could have been more marked for instance than her success in promoting Mrs. Beale's. She judged that if her whole history, for Mrs. Wix, had been the successive stages of her knowledge, so the very climax of the concatenation would, in the same view, be the stage at which the knowledge should overflow. As she was condemned to know more and more, how could it logically stop before she should know Most? It came to her in fact as they sat there on the sands that she was distinctly on the road to know Everything. She had not had governesses for nothing: what in the world had she ever done but learn and learn and learn? She looked at the pink sky with a placid foreboding that she soon should have learnt All. They lingered in the flushed air till at last it turned to grey and she seemed fairly to receive new information from every brush of the breeze. By the time they moved homeward it was as if this inevitability had become for Mrs. Wix a long, tense cord, twitched by a nervous hand, on which the valued pearls of intelligence were to be neatly strung.

In the evening upstairs they had another strange sensation, as to which Maisie could n't afterwards have told you whether it was bang in the middle or quite at the beginning that her companion sounded with fresh emphasis the note of the moral sense. What mattered was merely that she did exclaim, and again, as at first appeared, most disconnectedly: "God help me, it does seem to peep out!" Oh the queer confusions that had wooed it at last to such peeping! None so queer, however, as the words of woe, and it might verily be said of rage, in which the poor lady bewailed the tragic end of her own rich ignorance. There was a point at which she seized the child and hugged her as close as in the old days of partings and returns; at which she was visibly at a loss how to make up to such a victim for such contaminations: appealing, as to what she had done and was doing, in bewilderment, in explanation, in supplication, for reassurance, for pardon and even outright for pity.

"I don't know what I've said to you, my own: I don't know what I'm saying or what the turn you've given my life has rendered me, heaven forgive me, capable of saying. Have I lost all delicacy, all decency, all measure of how far and how bad? It seems to me mostly that I have, though I'm the last of whom you would ever have thought it. I've just done it for *you*, precious—not to lose you, which would have been worst of all: so that I've had to pay with my own innocence, if you do laugh! for clinging to you and keeping you. Don't let me pay for nothing; don't let me have been thrust for nothing into such horrors and such shames. I never knew anything about them and I never wanted to know! Now I know too much, too much!" the poor woman lamented and groaned. "I know so much that with hearing such talk I ask myself where I am; and with uttering it too, which is worse, say to myself that I'm far, too far, from where I started! I ask myself what I should have thought with my lost one if I had heard myself cross the line. There are lines I've crossed with *you* where I should have fancied I had come to a pretty pass—!" She gasped at the mere supposition. "I've gone from one thing to another, and all for the real love of you; and now what would any one say—I mean any one but *them*—if they were to hear the way I go on? I've had to keep up with you, have n't I?—and therefore what could I do less than look to you to keep up with *me*? But it's not *them* that are the worst—by which I mean to say it's not *him*: it's your dreadfully base papa and the one person in the world whom he could have found, I do believe —and she's not the Countess, duck—wickeder than himself. While they were about it at any rate, since they *were* ruining you, they might have done it so as to spare an honest woman. Then I should n't have had to do whatever it is that's the worst: throw up at you the badness you have n't taken in, or find my advantage in the vileness you *have*! What I did lose patience at this morning was at how it was that without your seeming to condemn—for you did n't, you remember!—you yet did seem to *know*. Thank God, in his mercy, at last, *if* you do!"

The night, this time, was warm and one of the windows stood open to the small balcony over the rail of which, on coming up from dinner, Maisie had hung a long time in the enjoyment of the chatter, the lights, the life of the quay made brilliant by the season and the hour. Mrs. Wix's requirements had drawn her in from this posture and Mrs. Wix's embrace had detained her even though midway in the outpouring her confusion and sympathy had permitted, or rather had positively helped, her to disengage herself. But the casement was still wide, the spectacle, the pleasure were still there, and from her place in the room, which, with its polished floor and its panels of elegance, was lighted from without more than from within, the child could still take account of them. She appeared to watch and listen; after which she answered Mrs. Wix with a question. "If I do know—?"

"If you do condemn." The correction was made with some austerity.

It had the effect of causing Maisie to heave a vague sigh of oppression and then after an instant and as if under cover of this ambiguity pass out again upon the balcony. She hung again over the rail; she felt the summer night; she dropped down into the manners of France. There was a café below the hotel, before which, with little chairs and tables, people sat on a space enclosed by plants in tubs; and the impression was enriched by the flash of the white aprons of waiters and the music of a man and a woman who, from beyond the precinct, sent up the strum of a guitar and the drawl of a song about "amour." Maisie knew what "amour" meant too, and wondered if Mrs. Wix did: Mrs. Wix remained within, as still as a mouse and perhaps not reached by the performance. After a while, but not till the musicians had ceased and begun to circulate with a little plate, her pupil came back to her. "*Is* it a crime?" Maisie then asked.

Mrs. Wix was as prompt as if she had been crouching in a lair. "Branded by the Bible."

"Well, he won't commit a crime."

Mrs. Wix looked at her gloomily. "He's committing one now."

"Now?"

"In being with her."

Maisie had it on her tongue's end to return once more: "But now he's free." She remembered, however, in time that one of the things she had known for the last entire hour was that this made no difference. After that, and as if to turn the right way, she was on the point of a blind dash, a weak reversion to the reminder that it might make a difference, might diminish the crime for Mrs. Beale; till such a reflexion was in its order also quashed by the visibility in Mrs. Wix's face of the collapse produced by her inference from her pupil's manner that after all her pains her pupil did n't even yet adequately understand. Never so much as when confronted had Maisie wanted to understand, and all her thought for a minute centred in the effort to come out with something which should be a disproof of her simplicity. "Just *trust* me, dear; that's all!"—she came out finally with that; and it was perhaps a good sign of her action that with a long, impartial moan Mrs. Wix floated her to bed.

There was no letter the next morning from Sir Claude—which Mrs. Wix let out that she deemed the worst of omens; yet it was just for the quieter communion they so got with him that, when after the coffee and rolls which made them more foreign than ever, it came to going forth for fresh drafts upon his credit they wandered again up the hill to the rampart instead of plunging into distraction with the crowd on the sands or into the sea with the semi-nude bathers. They gazed once more at their gilded Virgin; they sank once more upon their battered bench; they felt once more their distance from the Regent's Park. At last Mrs. Wix became definite about their friend's silence. "He *is* afraid of her! She has forbidden him to write." The fact of his fear Maisie already knew; but her companion's mention of it had at this moment two unexpected results. The first was her wondering in dumb remonstrance how Mrs. Wix, with a devotion not after all inferior to her own, could put into such an allusion such a grimness of derision; the second was that she found herself suddenly drop into a deeper view of it. She too had been afraid, as we have seen, of the people of whom Sir Claude was afraid, and by that law she had had her due measure of latent apprehension of Mrs. Beale. What occurred at present, however, was that, whereas this sympathy appeared vain as for him, the ground of it loomed dimly as a reason for selfish alarm. That uneasiness had not carried her far before Mrs. Wix spoke again and with an abruptness so great as almost to seem irrelevant. "Has it never occurred to you to be jealous of her?"

It never had in the least; yet the words were scarce in the air before Maisie had jumped at them. She held them well, she looked at them hard; at last she brought out with an assurance which there was no one, alas, but herself to admire: "Well, yes—since you ask me." She debated, then continued: "Lots of times!"

Mrs. Wix glared askance an instant; such approval as her look expressed was not wholly unqualified. It expressed at any rate something that presumably had to do with her saying once more: "Yes. He's afraid of her."

Maisie heard, and it had afresh its effect on her even through the blur of the attention now required by the possibility of that idea of jealousy—a possibility created only by her feeling she had thus found the way to show she was not simple. It struck out of Mrs. Wix that this lady still believed her moral sense to be interested and feigned; so what could be such a gage of her sincerity as a peep of the most restless of the passions? Such a revelation would baffle discouragement, and discouragement was in fact so baffled that, helped in some degree by the mere intensity of their need to hope, which also, according to its nature, sprang from the dark portent of the absent letter, the real pitch of their morning was reached by the note, not of mutual scrutiny, but of unprecedented frankness. There were broodings indeed and silences, and Maisie sank deeper into the vision that for her friend she was, at the most, superficial, and that also, positively, she was the more so the more she tried to appear complete. Was the sum of all knowledge only to know how little in this presence one would ever reach it? The answer to that question luckily lost itself in the brightness suffusing the scene as soon as Maisie had thrown out in regard to Mrs. Beale such a remark as she had never dreamed she should live to make. "If I thought she was unkind to him—I don't know *what* I should do!"

Mrs. Wix dropped one of her squints; she even confirmed it by a wild grunt. "I know what *I* should!"

Maisie at this felt that she lagged. "Well, I can think of *one* thing."

Mrs. Wix more directly challenged her. "What is it then?"

Maisie met her expression as if it were a game with forfeits for winking. "I'd *kill* her!" That at least, she hoped as she looked away, would guarantee her moral sense. She looked away, but her companion said nothing for so long that she at last turned her head again. Then she saw the straighteners all blurred with tears which after a little seemed to have sprung from her own eyes. There were tears in fact on both sides of the spectacles, and they were even so thick that it was presently all Maisie could do to make out through them that slowly, finally Mrs. Wix put forth a hand. It was the material pressure that settled this and even at the end of some minutes more things besides. It settled in its own way one thing in particular, which, though often, between them, heaven knew, hovered round and hung over, was yet to be established without the shadow of an attenuating smile. Oh there was no gleam of levity, as little of humour as of deprecation, in the long time they now sat together or in the way in which at some unmeasured point of it Mrs. Wix became distinct enough for her own dignity and yet not loud enough for the snoozing old women.

"I adore him. I adore him."

Maisie took it well in; so well that in a moment more she would have answered profoundly: "So do I." But before that moment passed something took place that brought other words to her lips; nothing more, very possibly, than the closer consciousness in her hand of the significance of Mrs. Wix's. Their hands remained linked in unutterable sign of their union, and what Maisie at last said was simply and serenely: "Oh I know!"

Their hands were so linked and their union was so confirmed that it took the far deep note of a bell, borne to them on the summer air, to call them back to a sense of hours and proprieties. They had touched bottom and melted together, but they gave a start at last: the bell was the voice of the inn and the inn was the image of luncheon. They should be late for it; they got up, and their quickened step on the return had something of the swing of confidence. When they reached the hotel the *table d'hôte* had begun; this was clear from the threshold, clear from the absence in the hall and on the stairs of the "personnel," as Mrs. Wix said—she had picked *that* up—all collected in the dining-room. They mounted to their apartments for a brush before the glass, and it was Maisie who, in passing and from a vain impulse, threw open the white and gold door. She was thus first to utter the sound that brought Mrs. Wix almost on top of her, as by the other accident it would have brought her on top of Mrs. Wix. It had at any rate the effect of leaving them bunched together in a strained stare at their new situation. This situation had put on in a flash the bright form of Mrs. Beale: she stood there in her hat and her jacket, amid bags and shawls, smiling and holding out her arms. If she had just arrived it was a different figure from either of the two that for *their* benefit, wan and tottering and none too soon to save life, the Channel had recently disgorged. She was as lovely as the day that had brought her over, as fresh as the luck and the health that attended her: it came to Maisie on the spot that she was more beautiful than she had ever been. All this was too quick to count, but there was still time in it to give the child the sense of what had kindled the light. That leaped out of the open arms, the open eyes, the open mouth; it leaped out with Mrs. Beale's loud cry at her: "I'm free, I'm free!"

XXVII

THE greatest wonder of all was the way Mrs. Beale addressed her announcement, so far as could be judged, equally to Mrs. Wix, who, as if from sudden failure of strength, sank into a chair while Maisie surrendered to the visitor's embrace. As soon as the child was liberated she met with profundity Mrs. Wix's stupefaction and actually was able to see that while in a manner sustaining the encounter her face yet seemed with intensity to say: "Now, for God's sake, don't crow 'I told you so!'" Maisie was somehow on the spot aware of an absence of disposition to crow; it had taken her but an extra minute to arrive at such a quick survey of the objects surrounding Mrs. Beale as showed that among them was no appurtenance of Sir Claude's. She knew his dressing-bag now—oh with the fondest knowledge!—and there was an instant during which its not being there was a stroke of the worst news. She was yet to learn what it could be to recognise in some lapse of a sequence the proof of an extinction, and therefore remained unaware that this momentary pang was a foretaste of the experience of death. It of course yielded in a flash to Mrs. Beale's brightness, it gasped itself away in her own instant appeal. "You've come alone?"

"Without Sir Claude?" Strangely, Mrs. Beale looked even brighter. "Yes; in the eagerness to get at you. You abominable little villain!"—and her stepmother, laughing clear, administered to her cheek a pat that was partly a pinch. "What were you up to and what did you take me for? But I'm glad to be abroad, and after all it's you who have shown me the way. I might n't, without you, have been able to come—to come, that is, so soon. Well, here I am at any rate and in a moment more I should have begun to worry about you. This will do very well"—she was good-natured about the place and even presently added that it was charming. Then with a rosier glow she made again her great point: "I'm free, I'm free!" Maisie made on her side her own: she carried back her gaze to Mrs. Wix, whom amazement continued to hold; she drew afresh her old friend's attention to the superior way she did n't take that up. What she did take up the next minute was the question of Sir Claude. "Where is he? Won't he come?"

Mrs. Beale's consideration of this oscillated with a smile between the two expectancies with which she was flanked: it was conspicuous, it was extraordinary, her unblinking acceptance of Mrs. Wix, a miracle of which Maisie had even now begun to read a reflexion in that lady's long visage. "He'll come, but we must *make* him!" she gaily brought forth.

"Make him?" Maisie echoed.

"We must give him time. We must play our cards."

"But he promised us awfully," Maisie replied.

"My dear child, he has promised *me* awfully; I mean lots of things, and not in every case kept his promise to the letter." Mrs. Beale's good humour insisted on taking for granted Mrs. Wix's, to whom her attention had suddenly grown prodigious. "I dare say he has done the same with you, and not always come to time. But he makes it up in his own way—and it is n't as if we did n't know exactly what he is. There's one thing he is," she went on, "which makes everything else only a question, for us, of tact." They scarce had time to wonder what this was before, as they might have said, it flew straight into their face. "He's as free as I am!"

"Yes, I know," said Maisie; as if, however, independently weighing the value of that. She really weighed also the oddity of her stepmother's treating it as news to *her*, who had been the first person literally to whom Sir Claude had mentioned it. For a few seconds, as if with the sound of it in her ears, she stood with him again, in memory and in the twilight, in the hotel garden at Folkestone.

Anything Mrs. Beale overlooked was, she indeed divined, but the effect of an exaltation of high spirits, a tendency to soar that showed even when she dropped—still quite impartially—almost to the confidential. "Well, then—we've only to wait. He can't do without us long. I'm sure, Mrs. Wix, he can't do without *you*! He's devoted to you; he has told me so much about you. The extent I count on you, you know, count on you to help me—!" was an extent that even all her radiance could n't express. What it could n't express quite as much as what it could made at any rate every instant her presence and even her famous freedom loom larger; and it was this mighty mass that once more led her companions, bewildered and disjoined, to exchange with each other as through a thickening veil confused and ineffectual signs. They clung together at least on the common ground of unpreparedness, and Maisie watched without relief the havoc of wonder in Mrs. Wix. It had reduced her to perfect impotence, and, but that gloom was black upon her, she sat as if fascinated by Mrs. Beale's high style. It had plunged her into a long deep hush; for what had happened was the thing she had least allowed for and before which the particular rigour she had worked up could only grow limp and sick. Sir Claude was to have reappeared with his accomplice or without her; never, never his accomplice without *him*. Mrs. Beale had gained apparently by this time an advantage she could pursue: she looked at the droll dumb figure with jesting reproach. "You really won't shake hands with me? Never mind; you'll come round!" She put the matter to no test, going on immediately and, instead of offering her hand, raising it, with a pretty gesture that her bent head met, to a long black pin that played a part in her back hair. "Are hats worn at luncheon? If you're as hungry as I am we must go right down."

Mrs. Wix stuck fast, but she met the question in a voice her pupil scarce recognised. "I wear mine."

Mrs. Beale, swallowing at one glance her brand-new bravery, which she appeared at once to refer to its origin and to follow in its flights, accepted this as conclusive. "Oh but I've not such a beauty!" Then she turned rejoicingly to Maisie. "I've got a beauty for *you*, my dear."

"A beauty?"

"A love of a hat—in my luggage. I remembered *that*"—she nodded at the object on her stepdaughter's head—"and I've brought you one with a peacock's breast. It's the most gorgeous blue!"

It was too strange, this talking with her there already not about Sir Claude but about peacocks—too strange for the child to have the presence of mind to thank her. But the felicity in which she had arrived was so proof against everything that Maisie felt more and more the depth of the purpose that must underlie it. She had a vague sense of its being abysmal, the spirit with which Mrs. Beale carried off the awkwardness, in the white and gold salon, of such a want of breath and of welcome. Mrs. Wix was more breathless than ever; the embarrassment of Mrs. Beale's isolation was as nothing to the embarrassment of her grace. The perception of this dilemma was the germ on the child's part of a new question altogether. What if *with* this indulgence—? But the idea lost itself in something too frightened for hope and too conjectured for fear; and while everything went by leaps and bounds one of the waiters stood at the door to remind them that the *table d'hôte* was half over.

"Had you come up to wash hands?" Mrs. Beale hereupon asked them. "Go and do it quickly and I'll be with you: they've put my boxes in that nice room—it was Sir Claude's. Trust him," she laughed, "to have a nice one!" The door of a neighbouring room stood open, and now from the threshold, addressing herself again to Mrs. Wix, she launched a note that gave the very key of what, as she would have said, she was up to. "Dear lady, please attend to my daughter."

She was up to a change of deportment so complete that it represented—oh for offices still honourably subordinate if not too explicitly menial—an absolute coercion, an interested clutch of the old woman's respectability. There was response, to Maisie's view, I may say at once, in the jump of that respectability to its feet: it was itself capable of one of the leaps, one of the bounds just mentioned, and it carried its charge, with this momentum and while Mrs. Beale popped into Sir Claude's chamber, straight away to where, at the end of the passage, pupil and governess were quartered. The greatest stride of all, for that matter, was that within a few seconds the pupil had, in another relation, been converted into a daughter. Maisie's eyes were still following it when, after the rush, with the door almost slammed and no thought of soap and towels, the pair stood face to face. Mrs. Wix, in this position, was the first to gasp a sound. "Can it ever be that *she* has one?"

Maisie felt still more bewildered. "One what?"

"Why moral sense."

They spoke as if you might have two, but Mrs. Wix looked as if it were not altogether a happy thought, and Maisie did n't see how even an affirmative from her own lips would clear up what had become most of a mystery. It was to this larger puzzle she sprang pretty straight. "*Is* she my mother now?"

It was a point as to which an horrific glimpse of the responsibility of an opinion appeared to affect Mrs. Wix like a blow in the stomach. She had evidently never thought of it; but she could think and rebound. "If she is, he's equally your father."

Maisie, however, thought further. "Then my father and my mother—!"

But she had already faltered and Mrs. Wix had already glared back: "Ought to live together? Don't begin it *again*!" She turned away with a groan, to reach the washing-stand, and Maisie could by this time recognise with a certain ease that that way verily madness did lie. Mrs. Wix gave a great untidy splash, but the next instant had faced round. "She has taken a new line."

"She was nice to you," Maisie concurred.

"What *she* thinks so—'go and dress the young lady!' But it's something!" she panted. Then she thought out the rest. "If he won't have her, why she'll have *you*. She'll be the one."

"The one to keep me abroad?"

"The one to give you a home." Mrs. Wix saw further; she mastered all the portents. "Oh she's cruelly clever! It's not a moral sense." She reached her climax. "It's a game!"

"A game?"

"Not to lose him. She has sacrificed him—to her duty."

"Then won't he come?" Maisie pleaded.

Mrs. Wix made no answer; her vision absorbed her. "He has fought. But she has won."

"Then won't he come?" the child repeated.

Mrs. Wix made it out. "Yes, hang him!" She had never been so profane.

For all Maisie minded! "Soon—to-morrow?"

"Too soon—whenever. Indecently soon."

"But then we *shall* be together!" the child went on. It

made Mrs. Wix look at her as if in exasperation; but nothing
had time to come before she precipitated: "Together with
you!" The air of criticism continued, but took voice only in
her companion's bidding her wash herself and come down.
The silence of quick ablutions fell upon them, presently
broken, however, by one of Maisie's sudden reversions.
"Mercy, is n't she handsome?"

Mrs. Wix had finished; she waited. "She'll attract attention." They were rapid, and it would have been noticed that
the shock the beauty had given them acted, incongruously,
as a positive spur to their preparations for rejoining her. She
had none the less, when they returned to the sitting-room,
already descended; the open door of her room showed it
empty and the chambermaid explained. Here again they were
delayed by another sharp thought of Mrs. Wix's. "But what
will she live on meanwhile?"

Maisie stopped short. "Till Sir Claude comes?"

It was nothing to the violence with which her friend had
been arrested. "Who'll pay the bills?"

Maisie thought. "Can't *she*?"

"She? She has n't a penny."

The child wondered. "But did n't papa—?"

"Leave her a fortune?" Mrs. Wix would have appeared to
speak of papa as dead had she not immediately added: "Why
he lives on other women!"

Oh yes, Maisie remembered. "Then can't he send—?"
She faltered again; even to herself it sounded queer.

"Some of their money to his wife?" Mrs. Wix gave a laugh
still stranger than the weird suggestion. "I dare say she'd
take it!"

They hurried on again; yet again, on the stairs, Maisie
pulled up. "Well, if she had stopped in England—!" she
threw out.

Mrs. Wix considered. "And he had come over instead?"

"Yes, as we expected." Maisie launched her speculation.
"What then would she have lived on?"

Mrs. Wix hung fire but an instant. "On other men!" And
she marched downstairs.

XXVIII

MRS. BEALE, at table between the pair, plainly attracted the
attention Mrs. Wix had foretold. No other lady present was
nearly so handsome, nor did the beauty of any other ac-
commodate itself with such art to the homage it produced.
She talked mainly to her other neighbour, and that left
Maisie leisure both to note the manner in which eyes were
riveted and nudges interchanged, and to lose herself in the
meanings that, dimly as yet and disconnectedly, but with
a vividness that fed apprehension, she could begin to read
into her stepmother's independent move. Mrs. Wix had
helped her by talking of a game; it was a connexion in which
the move could put on a strategic air. Her notions of diplomacy were thin, but it was a kind of cold diplomatic shoulder
and an elbow of more than usual point that, temporarily at
least, were presented to her by the averted inclination of
Mrs. Beale's head. There was a phrase familiar to Maisie,
so often was it used by this lady to express the idea of one's
getting what one wanted: one got it—Mrs. Beale always said
she at all events always got it or proposed to get it—by
"making love." She was at present making love, singular as
it appeared, to Mrs. Wix, and her young friend's mind had
never moved in such freedom as on thus finding itself face
to face with the question of what she wanted to get. This
period of the *omelette aux rognons* and the *poulet sauté*, while
her sole surviving parent, her fourth, fairly chattered to her
governess, left Maisie rather wondering if her governess
would hold out. It was strange, but she became on the spot
quite as interested in Mrs. Wix's moral sense as Mrs. Wix
could possibly be in hers: it had risen before her so pressingly
that this was something new for Mrs. Wix to resist. Resisting Mrs. Beale herself promised at such a rate to become
a very different business from resisting Sir Claude's view of
her. More might come of what had happened—whatever it
was—than Maisie felt she could have expected. She put it

together with a suspicion that, had she ever in her life had a sovereign changed, would have resembled an impression, baffled by the want of arithmetic, that her change was wrong: she groped about in it that she was perhaps playing the passive part in a case of violent substitution. A victim was what she should surely be if the issue between her step-parents had been settled by Mrs. Beale's saying: "Well, if she can live with but one of us alone, with which in the world should it be but me?" That answer was far from what, for days, she had nursed herself in, and the desolation of it was deepened by the absence of anything from Sir Claude to show he had not had to take it as triumphant. Had not Mrs. Beale, upstairs, as good as given out that she had quitted him with the snap of a tension, left him, dropped him in London, after some struggle as a sequel to which her own advent represented that she had practically sacrificed him? Maisie assisted in fancy at the probable episode in the Regent's Park, finding elements almost of terror in the suggestion that Sir Claude had not had fair play. They drew something, as she sat there, even from the pride of an association with such beauty as Mrs. Beale's; and the child quite forgot that, though the sacrifice of Mrs. Beale herself was a solution she had not invented, she would probably have seen Sir Claude embark upon it without a direct remonstrance.

What her stepmother had clearly now promised herself to wring from Mrs. Wix was an assent to the great modification, the change, as smart as a juggler's trick, in the interest of which nothing so much mattered as the new convenience of Mrs. Beale. Maisie could positively seize the moral that her elbow seemed to point in ribs thinly defended—the moral of its not mattering a straw which of the step-parents was the guardian. The essence of the question was that a girl was n't a boy: if Maisie had been a mere rough trousered thing, destined at the best probably to grow up a scamp, Sir Claude would have been welcome. As the case stood he had simply tumbled out of it, and Mrs. Wix would henceforth find herself in the employ of the right person. These arguments had really fallen into their place, for our young friend, at the very touch of that tone in which she had heard her new title declared. She was still, as a result of so many parents, a daughter to somebody even after papa and mamma were to all intents dead. If her father's wife and her mother's husband, by the operation of a natural or, for all she knew, a legal rule, were in the shoes of their defunct partners, then Mrs. Beale's partner was exactly as defunct as Sir Claude's and her shoes the very pair to which, in "Farange *v.* Farange and Others," the divorce court had given priority. The subject of that celebrated settlement saw the rest of her day really filled out with the pomp of all that Mrs. Beale assumed. The assumption rounded itself there between this lady's entertainers, flourished in a way that left them, in their bottomless element, scarce a free pair of eyes to exchange signals. It struck Maisie even a little that there was a rope or two Mrs. Wix might have thrown out if she would, a rocket or two she might have sent up. They had at any rate never been so long together without communion or telegraphy, and their companion kept them apart by simply keeping them with her. From this situation they saw the grandeur of their intenser relation to her pass and pass like an endless procession. It was a day of lively movement and of talk on Mrs. Beale's part so brilliant and overflowing as to represent music and banners. She took them out with her promptly to walk and to drive, and even—towards night—sketched a plan for carrying them to the Etablissement, where, for only a franc apiece, they should listen to a concert of celebrities. It reminded Maisie, the plan, of the side-shows at Earl's Court, and the franc sounded brighter than the shillings which had at that time failed; yet this too, like the other, was a frustrated hope: the francs failed like the shillings and the side-shows had set an example to the concert. The Etablissement in short melted away, and it was little wonder that a lady who from the moment of her arrival had been so gallantly in the breach should confess herself at last done up. Maisie could appreciate her fatigue; the day had not passed without such an observer's discovering that she was excited and even mentally comparing her state to that of the breakers after a gale. It had blown hard in London, and she would take time to go down. It was of the condition known to the child by report as that of talking against time that her emphasis, her spirit, her humour, which had never dropped, now gave the impression.

She too was delighted with foreign manners; but her daughter's opportunities of explaining them to her were un-expectedly forestalled by her own tone of large acquaintance with them. One of the things that nipped in the bud all response to her volubility was Maisie's surprised retreat before the fact that Continental life was what she had been almost brought up on. It was Mrs. Beale, disconcertingly, who began to explain it to her friends; it was she who, wherever they turned, was the interpreter, the historian and the guide. She was full of reference to her early travels—at the age of eighteen: she had at that period made, with a dis-tinguished Dutch family, a stay on the Lake of Geneva. Maisie had in the old days been regaled with anecdotes of these adventures, but they had with time become phantas-mal, and the heroine's quite showy exemption from be-wilderment at Boulogne, her acuteness on some of the very subjects on which Maisie had been acute to Mrs. Wix, were a high note of the majesty, of the variety of advantage, with which she had alighted. It was all a part of the wind in her sails and of the weight with which her daughter was now to feel her hand. The effect of it on Maisie was to add already the burden of time to her separation from Sir Claude. This might, to her sense, have lasted for days; it was as if, with their main agitation transferred thus to France and with neither mamma now nor Mrs. Beale nor Mrs. Wix nor her-self at his side, he must be fearfully alone in England. Hour after hour she felt as if she were waiting; yet she could n't have said exactly for what. There were moments when Mrs. Beale's flow of talk was a mere rattle to smother a knock. At no part of the crisis had the rattle so public a purpose as when, instead of letting Maisie go with Mrs. Wix to prepare for dinner, she pushed her—with a push at last incontestably maternal—straight into the room inherited from Sir Claude.

She titivated her little charge with her own brisk hands; then she brought out: "I'm going to divorce your father."

This was so different from anything Maisie had expected that it took some time to reach her mind. She was aware meanwhile that she probably looked rather wan. "To marry Sir Claude?"

Mrs. Beale rewarded her with a kiss. "It's sweet to hear you put it so."

This was a tribute, but it left Maisie balancing for an objection. "How *can* you when he's married?"

"He is n't—practically. He's free, you know."

"Free to marry?"

"Free, first, to divorce his own fiend."

The benefit that, these last days, she had felt she owed a certain person left Maisie a moment so ill-prepared for recognising this lurid label that she hesitated long enough to risk: "Mamma?"

"She is n't your mamma any longer," Mrs. Beale returned. "Sir Claude has paid her money to cease to be." Then as if remembering how little, to the child, a pecuniary transaction must represent: "She lets him off supporting her if he'll let her off supporting you."

Mrs. Beale appeared, however, to have done injustice to her daughter's financial grasp. "And support me himself?" Maisie asked.

"Take the whole bother and burden of you and never let her hear of you again. It's a regular signed contract."

"Why that's lovely of her!" Maisie cried.

"It's not so lovely, my dear, but that he'll get his divorce."

Maisie was briefly silent; after which, "No—he won't get it," she said. Then she added still more boldly: "And you won't get yours."

Mrs. Beale, who was at the dressing-glass, turned round with amusement and surprise. "How do you know that?"

"Oh I know!" cried Maisie.

"From Mrs. Wix?"

Maisie debated, then after an instant took her cue from Mrs. Beale's absence of anger, which struck her the more as she had felt how much of her courage she needed. "From Mrs. Wix," she admitted.

Mrs. Beale, at the glass again, made play with a powder-puff. "My own sweet, she's mistaken!" was all she said.

There was a certain force in the very amenity of this, but our young lady reflected long enough to remember that it was not the answer Sir Claude himself had made. The recollection nevertheless failed to prevent her saying: "Do you mean then that he won't come till he has got it?"

Mrs. Beale gave a last touch; she was ready; she stood there in all her elegance. "I mean, my dear, that it's because he *has* n't got it that I left him."

This opened a view that stretched further than Maisie could reach. She turned away from it, but she spoke before they went out again. "Do you like Mrs. Wix now?"

"Why, my chick, I was just going to ask you if you think she has come at all to like poor bad me!"

Maisie thought, at this hint; but unsuccessfully. "I have n't the least idea. But I'll find out."

"Do!" said Mrs. Beale, rustling out with her in a scented air and as if it would be a very particular favour.

The child tried promptly at bed-time, relieved now of the fear that their visitor would wish to separate her for the night from her attendant. "Have you held out?" she began as soon as the two doors at the end of the passage were again closed on them.

Mrs. Wix looked hard at the flame of the candle. "Held out—?"

"Why, she has been making love to you. Has she won you over?"

Mrs. Wix transferred her intensity to her pupil's face. "Over to what?"

"To *her* keeping me instead."

"Instead of Sir Claude?" Mrs. Wix was distinctly gaining time.

"Yes; who else? since it's not instead of you."

Mrs. Wix coloured at this lucidity. "Yes, that *is* what she means."

"Well, do you like it?" Maisie asked.

She actually had to wait, for oh her friend was embarrassed! "My opposition to the connexion—theirs—would then naturally to some extent fall. She has treated me to-day as if I were n't after all quite such a worm; not that I don't know very well where she got the pattern of her politeness. But of course," Mrs. Wix hastened to add, "I should n't like her as *the* one nearly so well as him."

"'Nearly so well!'" Maisie echoed. "I should hope indeed not." She spoke with a firmness under which she was herself the first to quiver. "I thought you 'adored' him."

"I do," Mrs. Wix sturdily allowed.

"Then have you suddenly begun to adore her too?"

Mrs. Wix, instead of directly answering, only blinked in support of her sturdiness. "My dear, in what a tone you ask that! You're coming out."

"Why should n't I? *You've* come out. Mrs. Beale has come out. We each have our turn!" And Maisie threw off the most extraordinary little laugh that had ever passed her young lips.

There passed Mrs. Wix's indeed the next moment a sound that more than matched it. "You're most remarkable!" she neighed.

Her pupil, though wholly without aspirations to pertness, barely faltered. "I think you've done a great deal to make me so."

"Very true, I have." She dropped to humility, as if she recalled her so recent self-arraignment.

"Would you accept her then? That's what I ask," said Maisie.

"As a substitute?" Mrs. Wix turned it over; she met again the child's eyes. "She has literally almost fawned upon me."

"She has n't fawned upon *him*. She has n't even been kind to him."

Mrs. Wix looked as if she had now an advantage. "Then do you propose to 'kill' her?"

"You don't answer my question," Maisie persisted. "I want to know if you accept her."

Mrs. Wix continued to hedge. "I want to know if *you* do!"

Everything in the child's person, at this, announced that it was easy to know. "Not for a moment."

"Not the two now?" Mrs. Wix had caught on; she flushed with it. "Only him alone?"

"Him alone or nobody."

"Not even *me*?" cried Mrs. Wix.

Maisie looked at her a moment, then began to undress. "Oh you're nobody!"

XXIX

HER sleep was drawn out; she instantly recognised lateness in the way her eyes opened to Mrs. Wix, erect, completely dressed, more dressed than ever, and gazing at her from the centre of the room. The next thing she was sitting straight up, wide awake with the fear of the hours of "abroad" that she might have lost. Mrs. Wix looked as if the day had already made itself felt, and the process of catching up with it began for Maisie in hearing her distinctly say: "My poor dear, he has come!"

"Sir Claude?" Maisie, clearing the little bed-rug with the width of her spring, felt the polished floor under her bare feet.

"He crossed in the night; he got in early." Mrs. Wix's head jerked stiffly backward. "He's there."

"And you've seen him?"

"No. He's there—he's there," Mrs. Wix repeated. Her voice came out with a queer extinction that was not a voluntary drop, and she trembled so that it added to their common emotion. Visibly pale, they gazed at each other.

"Isn't it too *beautiful*?" Maisie panted back at her; a challenge with an answer to which, however, she was not ready at once. The term Maisie had used was a flash of diplomacy— to prevent at any rate Mrs. Wix's using another. To that degree it was successful; there was only an appeal, strange and mute, in the white old face, which produced the effect of a want of decision greater than could by any stretch of optimism have been associated with her attitude toward what had happened. For Maisie herself indeed what had happened was oddly, as she could feel, less of a simple rapture than any arrival or return of the same supreme friend had ever been before. What had become overnight, what had become while she slept, of the comfortable faculty of gladness? She tried to wake it up a little wider by talking, by rejoicing, by plunging into water and into clothes, and she made out that it was ten o'clock, but also that Mrs. Wix had not yet breakfasted. The day before, at nine, they had had together a *café complet* in their sitting-room. Mrs. Wix on her side had evidently also a refuge to seek. She sought it in checking the precipitation of some of her pupil's present steps, in recalling to her with an approach to sternness that of such preliminaries those embodied in a thorough use of soap should be the most thorough, and in throwing even a certain reprobation on the idea of hurrying into clothes for the sake of a mere stepfather. She took her in hand with a silent insistence; she reduced the process to sequences more definite than any it had known since the days of Moddle. Whatever it might be that had now, with a difference, begun to belong to Sir Claude's presence was still after all compatible, for our young lady, with the instinct of dressing to see him with almost untidy haste. Mrs. Wix meanwhile luckily was not wholly directed to repression. "He's there—he's there!" she had said over several times. It was her answer to every invitation to mention how long she had been up and her motive for respecting so rigidly the slumber of her companion. It formed for some minutes her only account of the whereabouts of the others and her reason for not having yet seen them, as well as of the possibility of their presently being found in the salon.

"He's there—he's there!" she declared once more as she made, on the child, with an almost invidious tug, a strained undergarment "meet."

"Do you mean he's in the salon?" Maisie asked again.

"He's *with* her," Mrs. Wix desolately said. "He's with her," she reiterated.

"Do you mean in her own room?" Maisie continued.

She waited an instant. "God knows!"

Maisie wondered a little why, or how, God should know; this, however, delayed but an instant her bringing out: "Well, won't she go back?"

"Go back? Never!"

"She'll stay all the same?"

"All the more."

"Then won't Sir Claude go?" Maisie asked.

"Go back—if *she* doesn't?" Mrs. Wix appeared to give this question the benefit of a minute's thought. "Why should he have come—only to go back?"

Maisie produced an ingenious solution. "To *make* her go. To take her."

Mrs. Wix met it without a concession. "If he can make her go so easily, why should he have let her come?"

Maisie considered. "Oh just to see *me*. She has a right."

"Yes—she has a right."

"She's my mother!" Maisie tentatively tittered.

"Yes—she's your mother."

"Besides," Maisie went on, "he didn't let her come. He doesn't like her coming, and if he doesn't like it—"

Mrs. Wix took her up. "He must lump it—that's what he must do! Your mother was right about him—I mean your real one. He has no strength. No—none at all." She seemed more profoundly to muse. "He might have had some even with *her*—I mean with her ladyship. He's just a poor sunk slave," she asserted with sudden energy.

Maisie wondered again. "A slave?"

"To his passions."

She continued to wonder and even to be impressed; after which she went on: "But how do you know he'll stay?"

"Because he likes us!"—and Mrs. Wix, with her emphasis of the word, whirled her charge round again to deal with posterior hooks. She had positively never shaken her so. It was as if she quite shook something out of her. "But how will that help him if we—in spite of his liking!—don't stay?"

"Do you mean if we go off and leave him with her?"—Mrs. Wix put the question to the back of her pupil's head. "It *won't* help him. It will be his ruin. He'll have got nothing. He'll have lost everything. It will be his utter destruction, for he's certain after a while to loathe her."

"Then when he loathes her"—it was astonishing how she caught the idea—"he'll just come right after us!" Maisie announced.

"Never."

"Never?"

"She'll keep him. She'll hold him for ever."

Maisie doubted. "When he 'loathes' her?"

"That won't matter. She won't loathe *him*. People don't!" Mrs. Wix brought up.

"Some do. Mamma does," Maisie contended.

"Mamma does *not*!" It was startling—her friend contradicted her flat. "She loves him—she adores him. A woman knows." Mrs. Wix spoke not only as if Maisie were not a woman, but as if she would never be one. "*I* know!" she cried.

"Then why on earth has she left him?"

Mrs. Wix hesitated. "He hates *her*. Don't stoop so—lift up your hair. You know how I'm affected toward him," she added with dignity; "but you must also know that I see clear."

Maisie all this time was trying hard to do likewise. "Then if she has left him for that why shouldn't Mrs. Beale leave him?"

"Because she's not such a fool!"

"Not such a fool as mamma?"

"Precisely—if you *will* have it. Does it look like her leaving him?" Mrs. Wix enquired. She brooded again; then she went on with more intensity: "Do you want to know really and truly why? So that she may be his wretchedness and his punishment."

"His punishment?"—this was more than as yet Maisie could quite accept. "For what?"

"For everything. That's what will happen: he'll be tied to her for ever. She won't mind in the least his hating her, and she won't hate him back. She'll only hate *us*."

"Us?" the child faintly echoed.

"She'll hate *you*."

"Me? Why, I brought them together!" Maisie resentfully cried.

"You brought them together." There was a completeness in Mrs. Wix's assent. "Yes; it was a pretty job. Sit down." She began to brush her pupil's hair and, as she took up the mass of it with some force of hand, went on with a sharp recall: "Your mother adored him at first—it might have lasted. But he began too soon with Mrs. Beale. As you say," she pursued with a brisk application of the brush, "you brought them together."

"I brought them together"—Maisie was ready to reaffirm it. She felt none the less for a moment at the bottom of a hole; then she seemed to see a way out. "But I didn't bring mamma together—" She just faltered.

"With all those gentlemen?"—Mrs. Wix pulled her up. "No; it isn't quite so bad as that."

"I only said to the Captain"—Maisie had the quick memory of it—"that I hoped he at least (he was awfully nice!) would love her and keep her."

"And even that wasn't much harm," threw in Mrs. Wix.

"It wasn't much good," Maisie was obliged to recognise. "She can't bear him—not even a mite. She told me at Folkestone."

Mrs. Wix suppressed a gasp; then after a bridling instant during which she might have appeared to deflect with difficulty from her odd consideration of Ida's wrongs: "He was a nice sort of person for her to talk to you about!"

"Oh I *like* him!" Maisie promptly rejoined; and at this, with an inarticulate sound and an inconsequence still more marked, her companion bent over and dealt her on the cheek a rapid peck which had the apparent intention of a kiss.

"Well, if her ladyship doesn't agree with you, what does

it only prove?" Mrs. Wix demanded in conclusion. "It proves that she's fond of Sir Claude!"

Maisie, in the light of some of the evidence, reflected on that till her hair was finished, but when she at last started up she gave a sign of no very close embrace of it. She grasped at this moment Mrs. Wix's arm. "He must have got his divorce!"

"Since day before yesterday? Don't talk trash."

This was spoken with an impatience which left the child nothing to reply; whereupon she sought her defence in a completely different relation to the fact. "Well, I knew he would come!"

"So did I; but not in twenty-four hours. I gave him a few days!" Mrs. Wix wailed.

Maisie, whom she had now released, looked at her with interest. "How many did *she* give him?"

Mrs. Wix faced her a moment; then as if with a bewildered sniff: "You had better ask her!" But she had no sooner uttered the words than she caught herself up. "Lord o' mercy, how we talk!"

Maisie felt that however they talked she must see him, but she said nothing more for a time, a time during which she conscientiously finished dressing and Mrs. Wix also kept silence. It was as if they each had almost too much to think of, and even as if the child had the sense that her friend was watching her and seeing if she herself were watched. At last Mrs. Wix turned to the window and stood—sightlessly, as Maisie could guess—looking away. Then our young lady, before the glass, gave the supreme shake. "Well, I'm ready. And now to *see* him!"

Mrs. Wix turned round, but as if without having heard her. "It's tremendously grave." There were slow still tears behind the straighteners.

"It is—it is." Maisie spoke as if she were now dressed quite up to the occasion; as if indeed with the last touch she had put on the judgement-cap. "I must see him immediately."

"How can you see him if he does n't send for you?"

"Why can't I go and find him?"

"Because you don't know where he is."

"Can't I just look in the salon?" That still seemed simple to Maisie.

Mrs. Wix, however, instantly cut it off. "I would n't have you look in the salon for all the world!" Then she explained a little: "The salon is n't ours now."

"Ours?"

"Yours and mine. It's theirs."

"Theirs?" Maisie, with her stare, continued to echo. "You mean they want to keep us out?"

Mrs. Wix faltered; she sank into a chair and, as Maisie had often enough seen her do before, covered her face with her hands. "They ought to, at least. The situation's too monstrous!"

Maisie stood there a moment—she looked about the room. 'I 'll go to him—I 'll find him."

"*I* won't! I won't go *near* them!" cried Mrs. Wix.

"Then I 'll see him alone." The child spied what she had been looking for—she possessed herself of her hat. "Perhaps I 'll take him out!" And with decision she quitted the room.

When she entered the salon it was empty, but at the sound of the opened door some one stirred on the balcony, and Sir Claude, stepping straight in, stood before her. He was in light fresh clothes and wore a straw hat with a bright ribbon; these things, besides striking her in themselves as the very promise of the grandest of grand tours, gave him a certain radiance and, as it were, a tropical ease; but such an effect only marked rather more his having stopped short and, for a longer minute than had ever at such a juncture elapsed, not opened his arms to her. His pause made her pause and enabled her to reflect that he must have been up some time, for there were no traces of breakfast; and that though it was so late he had rather markedly not caused her to be called to him. Had Mrs. Wix been right about their forfeiture of the salon? Was it all his now, all his and Mrs. Beale's? Such an idea, at the rate her small thoughts throbbed, could only remind her of the way in which what had been hers hitherto was what was exactly most Mrs. Beale's and his. It was strange to be standing there and greeting him across a gulf, for he had by this time spoken, smiled and said: "My dear child, my dear child!" but without coming any nearer. In a flash she saw he was different—more so than he knew or designed. The next minute indeed it was as if he caught an impression from her face: this made him hold out his hand. Then they met, he kissed her, he laughed, she thought he even blushed: something of his affection rang out as usual. "Here I am, you see, again—as I promised you."

It was not as he had promised them—he had not promised them Mrs. Beale; but Maisie said nothing about that. What she said was simply: "I knew you had come. Mrs. Wix told me."

"Oh yes. And where is she?"

"In her room. She got me up—she dressed me."

Sir Claude looked at her up and down; a sweetness of mockery that she particularly loved came out in his face whenever he did that, and it was not wanting now. He raised his eyebrows and his arms to play at admiration; he was evidently after all disposed to be gay. "Got you up?— I should think so! She has dressed you most beautifully. Is n't she coming?"

Maisie wondered if she had better tell. "She said not."

"Does n't she want to see a poor devil?"

She looked about under the vibration of the way he described himself, and her eyes rested on the door of the room he had previously occupied. "Is Mrs. Beale in there?"

Sir Claude looked blankly at the same object. "I have n't the least idea!"

"You have n't seen her?"

"Not the tip of her nose."

Maisie thought: there settled on her, in the light of his beautiful smiling eyes, the faintest purest coldest conviction that he was n't telling the truth. "She has n't welcomed you?"

"Not by a single sign."

"Then where is she?"

236

Sir Claude laughed; he seemed both amused and surprised at the point she made of it. "I give it up!"

"Does n't she know you've come?"

He laughed again. "Perhaps she does n't care!"

Maisie, with an inspiration, pounced on his arm. "Has she *gone*?"

He met her eyes and then she could see that his own were really much graver than his manner. "Gone?" She had flown to the door, but before she could raise her hand to knock he was beside her and had caught it. "Let her be. I don't care about her. I want to see *you*."

Maisie fell back with him. "Then she *has n't* gone?"

He still looked as if it were a joke, but the more she saw of him the more she could make out that he was troubled. "It would n't be like her!"

She stood wondering at him. "Did you want her to come?"

"How can you suppose—?" He put it to her candidly. "We had an immense row over it."

"Do you mean you've quarrelled?"

Sir Claude was at a loss. "What has she told you?"

"That I'm hers as much as yours. That she represents papa."

His gaze struck away through the open window and up to the sky; she could hear him rattle in his trousers-pockets his money or his keys. "Yes—that's what she keeps saying." It gave him for a moment an air that was almost helpless.

"You say you don't care about her," Maisie went on. "*Do* you mean you've quarrelled?"

"We do nothing in life but quarrel."

He rose before her, as he said this, so soft and fair, so rich, in spite of what might worry him, in restored familiarities, that it gave a bright blur to the meaning—to what would otherwise perhaps have been the palpable promise—of the words. "Oh *your* quarrels!" she exclaimed with discouragement.

"I assure you hers are quite fearful!"

"I don't speak of hers. I speak of yours."

"Ah don't do it till I've had my coffee! You're growing

up clever," he added. Then he said: "I suppose you've breakfasted?"

"Oh no—I've had nothing."

"Nothing in your room?"—he was all compunction. "My dear old man!—we'll breakfast then together." He had one of his happy thoughts. "I say—we'll go out."

"That was just what I hoped. I've brought my hat."

"You *are* clever! We'll go to a café." Maisie was already at the door; he glanced round the room. "A moment—my stick." But there appeared to be no stick. "No matter; I left it—oh!" He remembered with an odd drop and came out.

"You left it in London?" she asked as they went downstairs.

"Yes—in London: fancy!"

"You were in such a hurry to come," Maisie explained.

He had his arm round her. "That must have been the reason." Halfway down he stopped short again, slapping his leg. "And poor Mrs. Wix?"

Maisie's face just showed a shadow. "Do you want her to come?"

"Dear no—I want to see you alone."

"That's the way I want to see *you*!" she replied. "Like before."

"Like before!" he gaily echoed. "But I mean has she had her coffee?"

"No, nothing."

"Then I'll send it up to her. Madame!" He had already, at the foot of the stair, called out to the stout *patronne*, a lady who turned to him from the bustling, breezy hall a countenance covered with fresh matutinal powder and a bosom as capacious as the velvet shelf of a chimneypiece, over which her round white face, framed in its golden frizzle, might have figured as a showy clock. He ordered, with particular recommendations, Mrs. Wix's repast, and it was a charm to hear his easy brilliant French: even his companion's ignorance could measure the perfection of it. The *patronne*, rubbing her hands and breaking in with high swift notes as into a florid duet, went with him to the street, and while they talked a moment longer Maisie remembered what Mrs. Wix had said about every one's liking him. It came out enough through the morning powder, it came out enough in the heaving bosom, how the landlady liked him. He had evidently ordered something lovely for Mrs. Wix. "*Et bien soigné, n'est-ce-pas?*"

"*Soyez tranquille*"—the *patronne* beamed upon him. "*Et pour Madame?*"

"*Madame?*" he echoed—it just pulled him up a little.

"*Rien encore?*"

"*Rien encore.* Come, Maisie." She hurried along with him, but on the way to the café he said nothing.

XXX

AFTER they were seated there it was different: the place was not below the hotel, but further along the quay; with wide, clear windows and a floor sprinkled with bran in a manner that gave it for Maisie something of the added charm of a circus. They had pretty much to themselves the painted spaces and the red plush benches; these were shared by a few scattered gentlemen who picked teeth, with facial contortions, behind little bare tables, and by an old personage in particular, a very old personage with a red ribbon in his buttonhole, whose manner of soaking buttered rolls in coffee and then disposing of them in the little that was left of the interval between his nose and chin might at a less anxious hour have cast upon Maisie an almost envious spell. They too had their *café au lait* and their buttered rolls, determined by Sir Claude's asking her if she could with that light aid wait till the hour of déjeuner. His allusion to this meal gave her, in the shaded sprinkled coolness, the scene, as she vaguely felt, of a sort of ordered mirrored licence, the haunt of those—the irregular, like herself—who went to bed or who rose too late, something to think over while she watched the white-aproned waiter perform as nimbly with plates and saucers as a certain conjurer her friend had in London taken her to a music-hall to see. Sir Claude had presently begun to talk again, to tell her how London had looked and how long he had felt himself, on either side, to have been absent; all about Susan Ash too and the amusement as well as the difficulty he had had with her; then all about his return journey and the Channel in the night and the crowd of people coming over and the way there were always too many one knew. He spoke of other matters beside, especially of what she must tell him of the occupations, while he was away, of Mrs. Wix and her pupil. Had n't they had the good time he had promised?—had he exaggerated a bit the arrangements made for their pleasure? Maisie had something—not all there was—to say of his success and of their gratitude: she had a complication of thought that grew every minute, grew with the consciousness that she had never seen him in this particular state in which he had been given back.

Mrs. Wix had once said—it was once or fifty times; once was enough for Maisie, but more was not too much—that he was wonderfully various. Well, he was certainly so, to the child's mind, on the present occasion: he was much more various than he was anything else. Besides, the fact that they were together in a shop, at a nice little intimate table as they had so often been in London, only made greater the difference of what they were together about. This difference was in his face, in his voice, in every look he gave her and every movement he made. They were not the looks and the movements he really wanted to show, and she could feel as well that they were not those she herself wanted. She had seen him nervous, she had seen every one she had come in contact with nervous, but she had never seen him so nervous as this. Little by little it gave her a settled terror, a terror that partook of the coldness she had felt just before, at the hotel, to find herself, on his answer about Mrs. Beale, disbelieve him. She seemed to see at present, to touch across the table, as if by laying her hand on it, what he had meant when he confessed on those several occasions to fear. Why was such a man so often afraid? It must have begun to come to her now that there was one thing just such a man above all could be afraid of. He could be afraid of himself. His fear at all events was there; his fear was sweet to her, beautiful and tender to her, was having coffee and buttered rolls and talk and laughter that were no talk and laughter at all with her; his fear was in his jesting postponing perverting voice; it was just in this make-believe way he had brought her out to imitate the old London playtimes, to imitate indeed a relation that had wholly changed, a relation that she had with her very eyes seen in the act of change when, the day before in the salon, Mrs. Beale rose suddenly before her. She rose before her, for that matter, now, and even while their refreshment delayed Maisie arrived at the straight question for which, on their entrance, his first word had given opportunity. "Are we going to have déjeuner with Mrs. Beale?"

His reply was anything but straight. "You and I?"

Maisie sat back in her chair. "Mrs. Wix and me."

Sir Claude also shifted. "That's an enquiry, my dear child, that Mrs. Beale herself must answer." Yes, he had shifted; but abruptly, after a moment during which something seemed to hang there between them and, as it heavily swayed, just fan them with the air of its motion, she felt that the whole thing was upon them. "Do you mind," he broke out, "my asking you what Mrs. Wix has said to you?"

"Said to me?"

"This day or two—while I was away."

"Do you mean about you and Mrs. Beale?"

Sir Claude, resting on his elbows, fixed his eyes a moment on the white marble beneath them. "No; I think we had a good deal of that—did n't we?—before I left you. It seems to me we had it pretty well all out. I mean about yourself, about your—don't you know?—associating with us, as I might say, and staying on with us. While you were alone with our friend what did she say?"

Maisie felt the weight of the question; it kept her silent for a space during which she looked at Sir Claude, whose eyes remained bent. "Nothing," she returned at last.

He showed incredulity. "Nothing?"

"Nothing," Maisie repeated; on which an interruption descended in the form of a tray bearing the preparations for their breakfast.

These preparations were as amusing as everything else; the waiter poured their coffee from a vessel like a watering-pot and then made it froth with the curved stream of hot milk that dropped from the height of his raised arm; but the two looked across at each other through the whole play of French pleasantness with a gravity that had now ceased to dissemble. Sir Claude sent the waiter off again for something and then took up her answer. "Has n't she tried to affect you?"

Face to face with him thus it seemed to Maisie that she had tried so little as to be scarce worth mentioning; again therefore an instant she shut herself up. Presently she found her middle course. "Mrs. Beale likes her now; and there 's one thing I 've found out—a great thing. Mrs. Wix enjoys her being so kind. She was tremendously kind all day yesterday."

"I see. And what did she do?" Sir Claude asked.

Maisie was now busy with her breakfast, and her companion attacked his own; so that it was all, in form at least, even more than their old sociability. "Everything she could think of. She was as nice to her as you are," the child said. "She talked to her all day."

"And what did she say to her?"

"Oh I don't know." Maisie was a little bewildered with his pressing her so for knowledge; it did n't fit into the degree of intimacy with Mrs. Beale that Mrs. Wix had so denounced and that, according to that lady, had now brought him back in bondage. Was n't he more aware than his stepdaughter of what would be done by the person to whom he was bound? In a moment, however, she added: "She made love to her."

Sir Claude looked at her harder, and it was clearly something in her tone that made him quickly say: "You don't mind my asking you, do you?"

"Not at all; only I should think you 'd know better than I."

"What Mrs. Beale did yesterday?"

She thought he coloured a trifle; but almost simultaneously with that impression she found herself answering: "Yes—if you *have* seen her."

He broke into the loudest of laughs. "Why, my dear boy, I told you just now I 've absolutely not. I say, don't you believe me?"

There was something she was already so afraid of that it covered up other fears. "Did n't you come back to see her?" she enquired in a moment. "Did n't you come back because you always want to so much?"

He received her enquiry as he had received her doubt— with an extraordinary absence of resentment. "I can imagine of course why you think that. But it does n't explain my doing what I have. It was, as I said to you just now at the inn, really and truly you I wanted to see."

She felt an instant as she used to feel when, in the back garden at her mother's, she took from him the highest push of a swing—high, high, high—that he had had put there for her pleasure and that had finally broken down under the weight and the extravagant patronage of the cook. "Well, that 's beautiful. But to see me, you mean, and go away again?"

"My going away again is just the point. I can't tell yet— it all depends."

"On Mrs. Beale?" Maisie asked. "*She* won't go away." He finished emptying his coffee-cup and then, when he had put it down, leaned back in his chair, where she could see that he smiled on her. This only added to her idea that he was in trouble, that he was turning somehow in his pain and trying different things. He continued to smile and she went on: "Don't you know that?"

"Yes, I may as well confess to you that as much as that I do know. *She* won't go away. She'll stay."

"She'll stay. She'll stay," Maisie repeated.

"Just so. Won't you have some more coffee?"

"Yes, please."

"And another buttered roll?"

"Yes, please."

243

He signed to the hovering waiter, who arrived with the shining spout of plenty in either hand and with the friendliest interest in mademoiselle. "*Les tartines sont là.*" Their cups were replenished and, while he watched almost musingly the bubbles in the fragrant mixture, "Just so—just so," Sir Claude said again and again. "It's awfully awkward!" he exclaimed when the waiter had gone.

"That she won't go?"

"Well—everything! Well, well, well!" But he pulled himself together; he began again to eat. "I came back to ask you something. That's what I came back for."

"I know what you want to ask me," Maisie said.

"Are you very sure?"

"I'm *almost* very."

"Well then risk it. You must n't make *me* risk everything."

She was struck with the force of this. "You want to know if I should be happy with *them*."

"With those two ladies only? No, no, old man: *vous n'y êtes pas.* So now—there!" Sir Claude laughed.

"Well then what is it?"

The next minute, instead of telling her what it was, he laid his hand across the table on her own and held her as if under the prompting of a thought. "Mrs. Wix would stay with *her*?"

"Without you? Oh yes—now."

"On account, as you just intimated, of Mrs. Beale's changed manner?"

Maisie, with her sense of responsibility, weighed both Mrs. Beale's changed manner and Mrs. Wix's human weakness. "I think she talked her round."

Sir Claude thought a moment. "Ah poor dear!"

"Do you mean Mrs. Beale?"

"Oh no—Mrs. Wix."

"She likes being talked round—treated like any one else. Oh she likes great politeness," Maisie expatiated. "It affects her very much."

Sir Claude, to her surprise, demurred a little to this. "Very much—up to a certain point."

"Oh up to any point!" Maisie returned with emphasis.

"Well, have n't I been polite to her?"

"Lovely—and she perfectly worships you."

"Then, my dear child, why can't she let me alone?"—this time Sir Claude unmistakeably blushed. Before Maisie, however, could answer his question, which would indeed have taken her long, he went on in another tone: "Mrs. Beale thinks she has probably quite broken her down. But she has n't."

Though he spoke as if he were sure, Maisie was strong in the impression she had just uttered and that she now again produced. "She has talked her round."

"Ah yes; round to herself, but not round to me."

Oh she could n't bear to hear him say that! "To you? Don't you really believe how she loves you?"

Sir Claude examined his belief. "Of course I know she's wonderful."

"She's just every bit as fond of you as *I* am," said Maisie. "She told me so yesterday."

"Ah then," he promptly exclaimed, "she *has* tried to affect you! I don't love *her*, don't you see? I do her perfect justice," he pursued, "but I mean I don't love her as I do you, and I'm sure you would n't seriously expect it. She's not my daughter—come, old chap! She's not even my mother, though I dare say it would have been better for me if she had been. I'll do for her what I'd do for my mother, but I won't do more." His real excitement broke out in a need to explain and justify himself, though he kept trying to correct and conceal it with laughs and mouthfuls and other vain familiarities. Suddenly he broke off, wiping his moustache with sharp pulls and coming back to Mrs. Beale. "Did she try to talk *you* over?"

"No—to me she said very little. Very little indeed," Maisie continued.

Sir Claude seemed struck with this. "She was only sweet to Mrs. Wix?"

"As sweet as sugar!" cried Maisie.

He looked amused at her comparison, but he did n't

contest it; he uttered on the contrary, in an assenting way, a little inarticulate sound. "I know what she *can* be. But much good may it have done her! Mrs. Wix won't *come* 'round.' That's what makes it so fearfully awkward."

Maisie knew it was fearfully awkward; she had known this now, she felt, for some time, and there was something else it more pressingly concerned her to learn. "What is it you meant you came over to ask me?"

"Well," said Sir Claude, "I was just going to say. Let me tell you it will surprise you." She had finished breakfast now and she sat back in her chair again: she waited in silence to hear. He had pushed the things before him a little way and had his elbows on the table. This time, she was convinced, she knew what was coming, and once more, for the crash, as with Mrs. Wix lately in her room, she held her breath and drew together her eyelids. He was going to say she must give him up. He looked hard at her again; then he made his effort. "Should you see your way to let her go?"

She was bewildered. "To let who—?"

"Mrs. Wix simply. I put it at the worst. Should you see your way to sacrifice her? Of course I know what I'm asking."

Maisie's eyes opened wide again; this was so different from what she had expected. "And stay with you alone?"

He gave another push to his coffee-cup. "With me and Mrs. Beale. Of course it would be rather rum; but everything in our whole story is rather rum, you know. What's more unusual than for any one to be given up, like you, by her parents?"

"Oh nothing is more unusual than *that*!" Maisie concurred, relieved at the contact of a proposition as to which concurrence could have lucidity.

"Of course it would be quite unconventional," Sir Claude went on—"I mean the little household we three should make together; but things have got beyond that, don't you see? They got beyond that long ago. We shall stay abroad at any rate—it's ever so much easier and it's our affair and nobody else's: it's no one's business but ours on all the blessed earth. I don't say that for Mrs. Wix, poor dear—I do her absolute justice. I respect her; I see what she means; she has done me a lot of good. But there are the facts. There they are, simply. And here am I, and here are you. And she won't come round. She's right from her point of view. I'm talking to you in the most extraordinary way—I'm always talking to you in the most extraordinary way, ain't I? One would think you were about sixty and that I—I don't know what any one would think *I* am. Unless a beastly cad!" he suggested. "I've been awfully worried, and this's what it has come to. You've done us the most tremendous good, and you'll do it still and always, don't you see? We can't let you go—you're everything. There are the facts as I say. She *is* your mother now, Mrs. Beale, by what has happened, and I, in the same way, I'm your father. No one can contradict that, and we can't get out of it. My idea would be a nice little place—somewhere in the South—where she and you would be together and as good as any one else. And I should be as good too, don't you see? for I should n't live with you, but I should be close to you—just round the corner, and it would be just the same. My idea would be that it should all be perfectly open and frank. *Honi soit qui mal y pense*, don't you know? You're the best thing—you and what we can do for you—that either of us has ever known:" he came back to that. "When I say to her 'Give her up, come,' she lets me have it bang in the face: 'Give her up yourself!' It's the same old vicious circle—and when I say vicious I don't mean a pun, a what-d'-ye-call-'em. Mrs. Wix is the obstacle; I mean, you know, if she has affected you. She has affected *me*, and yet here I am. I never was in such a tight place: please believe it's only that that makes me put it to you as I do. My dear child, is n't that—to put it so—just the way out of it? That came to me yesterday, in London, after Mrs. Beale had gone: I had the most infernal atrocious day. 'Go straight over and put it to her: let her choose, freely, her own self.' So I do, old girl—I put it to you. *Can* you choose freely?"

This long address, slowly and brokenly uttered, with fidgets and falterings, with lapses and recoveries, will a mottled face and embarrassed but supplicating eyes, reached the child from a quarter so close that after the shock of the first sharpness she could see intensely its direction and follow it from point to point; all the more that it came back to the point at which it had started. There was a word that had hummed all through it. "Do you call it a 'sacrifice'?"

"Of Mrs. Wix? I'll call it whatever *you* call it. I won't funk it—I have n't, have I? I'll face it in all its baseness. Does it strike you it *is* base for me to get you well away from her, to smuggle you off here into a corner and bribe you with sophistries and buttered rolls to betray her?"

"To betray her?"

"Well—to part with her."

Maisie let the question wait; the concrete image it presented was the most vivid side of it. "If I part with her where will she go?"

"Back to London."

"But I mean what will she do?"

"Oh as for that I won't pretend I know. I don't. We all have our difficulties."

That, to Maisie, was at this moment more striking than it had ever been. "Then who'll teach me?"

Sir Claude laughed out. "What Mrs. Wix teaches?"

She smiled dimly; she saw what he meant. "It is n't so very very much."

"It's so very very little," he returned, "that that's a thing we've positively to consider. We probably should n't give you another governess. To begin with we should n't be able to get one—not of the only kind that would do. It would n't do—the kind that *would* do," he queerly enough explained. "I mean they would n't stay—heigh-ho! We'd do you ourselves. Particularly me. You see I *can* now; I have n't got to mind—what I used to. I won't fight shy as I did—she can show out *with* me. Our relation, all round, is more regular."

It seemed wonderfully regular, the way he put it; yet none the less, while she looked at it as judiciously as she could, the picture it made persisted somehow in being a combination quite distinct—an old woman and a little girl seated in deep silence on a battered old bench by the rampart of the *haute ville*. It was just at that hour yesterday; they were hand in hand; they had melted together. "I don't think you yet understand how she clings to you," Maisie said at last.

"I do—I do. But for all that—!" And he gave, turning in his conscious exposure, an oppressed impatient sigh; the sigh, even his companion could recognise, of the man naturally accustomed to that argument, the man who wanted thoroughly to be reasonable, but who, if really he had to mind so many things, would be always impossibly hampered. What it came to indeed was that he understood quite perfectly. If Mrs. Wix clung it was all the more reason for shaking Mrs. Wix off.

This vision of what she had brought him to occupied our young lady while, to ask what he owed, he called the waiter and put down a gold piece that the man carried off for change. Sir Claude looked after him, then went on: "How could a woman have less to reproach a fellow with? I mean as regards herself."

Maisie entertained the question. "Yes. How *could* she have less? So why are you so sure she'll go?"

"Surely you heard why—you heard her come out three nights ago? How can she do anything but go—after what she then said? I've done what she warned me of—she was absolutely right. So here we are. Her liking Mrs. Beale, as you call it now, is a motive sufficient, with other things, to make her, for your sake, stay on without me; it's not a motive sufficient to make her, even for yours, stay on *with* me—swallow, don't you see? what she can't swallow. And when you say she's as fond of me as you are I think I can, if that's the case, challenge you a little on it. Would *you*, only with those two, stay on without me?" The waiter came back with the change, and that gave her, under this appeal, a moment's respite. But when he had retreated again with the "tip" gathered in with graceful thanks on a subtle hint from Sir Claude's forefinger, the latter, while pocketing the money, followed the appeal up. "Would you let her make you live with Mrs. Beale?"

"Without you? Never," Maisie then answered. "Never," she said again.

It made him quite triumph, and she was indeed herself shaken by the mere sound of it. "So you see you're not, like her," he exclaimed, "so ready to give me away!" Then he came back to his original question. "*Can* you choose? I mean can you settle it by a word yourself? Will you stay on with us without her?"

Now in truth she felt the coldness of her terror, and it seemed to her that suddenly she knew, as she knew it about Sir Claude, what she was afraid of. She was afraid of herself. She looked at him in such a way that it brought, she could see, wonder into his face, a wonder held in check, however, by his frank pretension to play fair with her, not to use advantages, not to hurry nor hustle her—only to put her chance clearly and kindly before her. "May I think?" she finally asked.

"Certainly, certainly. But how long?"

"Oh only a little while," she said meekly.

He had for a moment the air of wishing to look at it as if it were the most cheerful prospect in the world. "But what shall we do while you're thinking?" He spoke as if thought were compatible with almost any distraction.

There was but one thing Maisie wished to do, and after an instant she expressed it. "Have we got to go back to the hotel?"

"Do you want to?"

"Oh no."

"There's not the least necessity for it." He bent his eyes on his watch; his face was now very grave. "We can do anything else in the world." He looked at her again almost as if he were on the point of saying that they might for instance start off for Paris. But even while she wondered if that were not coming he had a sudden drop. "We can take a walk."

She was all ready, but he sat there as if he had still something more to say. This too, however, did n't come; so she herself spoke. "I think I should like to see Mrs. Wix first."

"Before you decide? All right—all right." He had put on

his hat, but he had still to light a cigarette. He smoked a minute, with his head thrown back, looking at the ceiling; then he said: "There's one thing to remember—I've a right to impress it on you: we stand absolutely in the place of your parents. It's their defection, their extraordinary baseness, that has made our responsibility. Never was a young person more directly committed and confided." He appeared to say this over, at the ceiling, through his smoke, a little for his own illumination. It carried him after a pause somewhat further. "Though I admit it was to each of us separately."

He gave her so at that moment and in that attitude the sense of wanting, as it were, to be on her side—on the side of what would be in every way most right and wise and charming for her—that she felt a sudden desire to prove herself not less delicate and magnanimous, not less solicitous for his own interests. What were these but that of the "regularity" he had just before spoken of? "It *was* to each of you separately," she accordingly with much earnestness remarked. "But don't you remember? I brought you together."

He jumped up with a delighted laugh. "Remember? Rather! You brought us together, you brought us together. Come!"

XXXI

SHE remained out with him for a time of which she could take no measure save that it was too short for what she wished to make of it—an interval, a barrier indefinite, insurmountable. They walked about, they dawdled, they looked in shop-windows; they did all the old things exactly as if to try to get back all the old safety, to get something out of them that they had always got before. This had come before, whatever it was, without their trying, and nothing came now but the intenser consciousness of their quest and their subterfuge. The strangest thing of all was what had really happened to the old safety. What had really happened was that Sir Claude was "free" and that Mrs. Beale was "free," and yet that the new medium was somehow still more oppressive than the old. She could feel that Sir Claude concurred with her in the sense that the oppression would be worst at the inn, where, till something should be settled, they would feel the want of something—of what could they call it but a footing? The question of the settlement loomed larger to her now: it depended, she had learned, so completely on herself. Her choice, as her friend had called it, was there before her like an impossible sum on a slate, a sum that in spite of her plea for consideration she simply got off from doing while she walked about with him. She must see Mrs. Wix before she could do her sum; therefore the longer before she saw her the more distant would be the ordeal. She met at present no demand whatever of her obligation; she simply plunged, to avoid it, deeper into the company of Sir Claude. She saw nothing that she had seen hitherto—no touch in the foreign picture that had at first been always before her. The only touch was that of Sir Claude's hand, and to feel her own in it was her mute resistance to time. She went about as sightlessly as if he had been leading her blindfold. If they were afraid of themselves it was themselves they would find at the inn. She was certain now that what awaited them there would be to lunch with Mrs. Beale. All her instinct was to avoid that, to draw out their walk, to find pretexts, to take him down upon the beach, to take him to the end of the pier. He said no other word to her about what they had talked of at breakfast, and she had a dim vision of how his way of not letting her see him definitely wait for anything from her would make any one who should know of it, would make Mrs. Wix for instance, think him more than ever a gentleman. It was true that once or twice, on the jetty, on the sands, he looked at her for a minute with eyes that seemed to propose to her to come straight off with him to Paris. That, however, was not to give her a nudge about her responsibility. He evidently wanted to procrastinate quite as much as she did; he was not a bit more in a hurry to get back to the others. Maisie herself at this moment could be secretly merciless to Mrs. Wix—to the extent at any rate of not caring if her continued disappearance did make that lady begin to worry about what had become of her, even begin to wonder perhaps if the truants had n't found their remedy. Her want of mercy to Mrs. Beale indeed was at least as great; for Mrs. Beale's worry and wonder would be as much greater as the object at which they were directed. When at last Sir Claude, at the far end of the *plage*, which they had already, in the many-coloured crowd, once traversed, suddenly, with a look at his watch, remarked that it was time, not to get back to the *table d'hôte*, but to get over to the station and meet the Paris papers—when he did this she found herself thinking quite with intensity what Mrs. Beale and Mrs. Wix *would* say. On the way over to the station she had even a mental picture of the stepfather and the pupil established in a little place in the South while the governess and the stepmother, in a little place in the North, remained linked by a community of blankness and by the endless series of remarks it would give birth to. The Paris papers had come in and her companion, with a strange extravagance, purchased no fewer than eleven: it took up time while they hovered at the bookstall on the restless platform, where the little volumes in a row were all yellow and pink and one of her favourite old women in one of her favourite old caps absolutely wheedled him into the purchase of three. They had thus so much to carry home that it would have seemed simpler, with such a provision for a nice straight journey through France, just to "nip," as she phrased it to herself, into the coupé of the train that, a little further along, stood waiting to start. She asked Sir Claude where it was going.

"To Paris. Fancy!"

She could fancy well enough. They stood there and smiled, he with all the newspapers under his arm and she with the three books, one yellow and two pink. He had told her the pink were for herself and the yellow one for Mrs. Beale, implying in an interesting way that these were the natural divisions in France of literature for the young and for the old. She knew how prepared they looked to pass into the train, and she presently brought out to her companion: "I wish we could go. Won't you take me?"

He continued to smile. "Would you really come?"

"Oh yes, oh yes. Try."

"Do you want me to take our tickets?"

"Yes, take them."

"Without any luggage?"

She showed their two armfuls, smiling at him as he smiled at her, but so conscious of being more frightened than she had ever been in her life that she seemed to see her whiteness as in a glass. Then she knew that what she saw was Sir Claude's whiteness: he was as frightened as herself. "Have n't we got plenty of luggage?" she asked. "Take the tickets— have n't you time? When does the train go?"

Sir Claude turned to a porter. "When does the train go?"

The man looked up at the station-clock. "In two minutes. *Monsieur est placé?*"

"*Pas encore.*"

"*Et vos billets?—vous n'avez que le temps.*" Then after a look at Maisie, "*Monsieur veut-il que je les prenne?*" the man said.

Sir Claude turned back to her. "*Veux-tu bien qu'il en prenne?*"

It was the most extraordinary thing in the world: in the intensity of her excitement she not only by illumination understood all their French, but fell into it with an active perfection. She addressed herself straight to the porter. "*Prenny, prenny. Oh prenny!*"

"*Ah si mademoiselle le veut—!*" He waited there for the money.

But Sir Claude only stared—stared at her with his white face. "You *have* chosen then? You'll let her go?"

Maisie carried her eyes wistfully to the train, where, amid cries of "*En voiture, en voiture!*" heads were at windows and doors banging loud. The porter was pressing. "*Ah vous n'avez plus le temps!*"

"It's going—it's going!" cried Maisie.

They watched it move, they watched it start; then the

man went his way with a shrug. "It's gone!" Sir Claude said.

Maisie crept some distance up the platform; she stood there with her back to her companion, following it with her eyes, keeping down tears, nursing her pink and yellow books. She had had a real fright but had fallen back to earth. The odd thing was that in her fall her fear too had been dashed down and broken. It was gone. She looked round at last, from where she had paused, at Sir Claude's, and then saw that his was n't. It sat there with him on the bench to which, against the wall of the station, he had retreated, and where, leaning back and, as she thought, rather queer, he still waited. She came down to him and he continued to offer his ineffectual intention of pleasantry. "Yes, I've chosen," she said to him. "I'll let her go if you—if you—"

She faltered; he quickly took her up. "If I, if I—?"

"If you'll give up Mrs. Beale."

"Oh!" he exclaimed; on which she saw how much, how hopelessly he was afraid. She had supposed at the café that it was of his rebellion, of his gathering motive; but how could that be when his temptations—that temptation for example of the train they had just lost—were after all so slight? Mrs. Wix was right. He was afraid of his weakness—of his weakness.

She could n't have told you afterwards how they got back to the inn: she could only have told you that even from this point they had not gone straight, but once more had wandered and loitered and, in the course of it, had found themselves on the edge of the quay where—still apparently with half an hour to spare—the boat prepared for Folkestone was drawn up. Here they hovered as they had done at the station; here they exchanged silences again, but only exchanged silences. There were punctual people on the deck, choosing places, taking the best; some of them already contented, all established and shawled, facing to England and attended by the steward, who, confined on such a day to the lighter offices, tucked up the ladies' feet or opened bottles with a pop. They looked down at these things without a word; they even picked out a good place for two that was left in the lee of a lifeboat; and if they lingered rather stupidly, neither deciding to go aboard nor deciding to come away, it was Sir Claude quite as much as she who would n't move. It was Sir Claude who cultivated the supreme stillness by which she knew best what he meant. He simply meant that he knew all she herself meant. But there was no pretence of pleasantry now: their faces were grave and tired. When at last they lounged off it was as if his fear, his fear of his weakness, leaned upon her heavily as they followed the harbour. In the hall of the hotel as they passed in she saw a battered old box that she recognised, an ancient receptacle with dangling labels that she knew and a big painted W, lately done over and intensely personal, that seemed to stare at her with a recognition and even with some suspicion of its own. Sir Claude caught it too, and there was agitation for both of them in the sight of this object on the move. Was Mrs. Wix going and was the responsibility of giving her up lifted, at a touch, from her pupil? Her pupil and her pupil's companion, transfixed a moment, held, in the presence of the omen, communication more intense than in the presence either of the Paris train or of the Channel steamer; then, and still without a word, they went straight upstairs. There, however, on the landing, out of sight of the people below, they collapsed so that they had to sink down together for support: they simply seated themselves on the uppermost step while Sir Claude grasped the hand of his stepdaughter with a pressure that at another moment would probably have made her squeal. Their books and papers were all scattered. "She thinks you 've given her up!"

"Then I must see her—I must see her," Maisie said.

"To bid her good-bye?"

"I must see her—I must see her," the child only repeated.

They sat a minute longer, Sir Claude, with his tight grip of her hand and looking away from her, looking straight down the staircase to where, round the turn, electric bells rattled and the pleasant sea-draught blew. At last, loosening his grasp, he slowly got up while she did the same. They went together along the lobby, but before they reached the salon he stopped again. "If I give up Mrs. Beale—?"

"I'll go straight out with you again and not come back till she has gone."

He seemed to wonder. "Till Mrs. Beale—?"

He had made it sound like a bad joke. "I mean till Mrs. Wix leaves—in that boat."

Sir Claude looked almost foolish. "Is she going in that boat?"

"I suppose so. I won't even bid her good-bye," Maisie continued. "I'll stay out till the boat has gone. I'll go up to the old rampart."

"The old rampart?"

"I'll sit on that old bench where you see the gold Virgin."

"The gold Virgin?" he vaguely echoed. But it brought his eyes back to her as if after an instant he could see the place and the thing she named—could see her sitting there alone. "While I break with Mrs. Beale?"

"While you break with Mrs. Beale."

He gave a long deep smothered sigh. "I must see her first."

"You won't do as I do? Go out and wait?"

"Wait?"—once more he appeared at a loss.

"Till they both have gone," Maisie said.

"Giving *us* up?"

"Giving *us* up."

Oh with what a face for an instant he wondered if that could be! But his wonder the next moment only made him go to the door and, with his hand on the knob, stand as if listening for voices. Maisie listened, but she heard none. All she heard presently was Sir Claude's saying with speculation quite choked off, but so as not to be heard in the salon: "Mrs. Beale will never go." On this he pushed open the door and she went in with him. The salon was empty, but as an effect of their entrance the lady he had just mentioned appeared at the door of the bedroom. "Is she going?" he then demanded.

Mrs. Beale came forward, closing her door behind her. "I've had the most extraordinary scene with her. She told me yesterday she'd stay."

"And my arrival has altered it?"

"Oh we took that into account!" Mrs. Beale was flushed, which was never quite becoming to her, and her face visibly testified to the encounter to which she alluded. Evidently, however, she had not been worsted, and she held up her head and smiled and rubbed her hands as if in sudden emulation of the *patronne*. "She promised she'd stay even if you should come."

"Then why has she changed?"

"Because she's a hound. The reason she herself gives is that you've been out too long."

Sir Claude stared. "What has that to do with it?"

"You've been out an age," Mrs. Beale continued; "I myself could n't imagine what had become of you. The whole morning," she exclaimed, "and luncheon long since over!"

Sir Claude appeared indifferent to that. "Did Mrs. Wix go down with you?" he only asked.

"Not she; she never budged!"—and Mrs. Beale's flush, to Maisie's vision, deepened. "She moped there—she did n't so much as come out to me; and when I sent to invite her she simply declined to appear. She said she wanted nothing, and I went down alone. But when I came up, fortunately a little primed"—and Mrs. Beale smiled a fine smile of battle—"she *was* in the field!"

"And you had a big row?"

"We had a big row"—she assented with a frankness as large. "And while you left me to that sort of thing I should like to know where you were!" She paused for a reply, but Sir Claude merely looked at Maisie; a movement that promptly quickened her challenge. "Where the mischief have you been?"

"You seem to take it as hard as Mrs. Wix," Sir Claude returned.

"I take it as I choose to take it, and you don't answer my question."

He looked again at Maisie—as if for an aid to this effort; whereupon she smiled at her stepmother and offered: "We've been everywhere."

Mrs. Beale, however, made her no response, thereby adding to a surprise of which our young lady had already felt the light brush. She had received neither a greeting nor a glance, but perhaps this was not more remarkable than the omission, in respect to Sir Claude, parted with in London two days before, of any sign of a sense of their reunion. Most remarkable of all was Mrs. Beale's announcement of the pledge given by Mrs. Wix and not hitherto revealed to her pupil. Instead of heeding this witness she went on with acerbity: "It might surely have occurred to you that something would come up."

Sir Claude looked at his watch. "I had no idea it was so late, nor that we had been out so long. We were n't hungry. It passed like a flash. What *has* come up?"

"Oh that she's disgusted," said Mrs. Beale.

"With whom then?"

"With Maisie." Even now she never looked at the child, who stood there equally associated and disconnected. "For having no moral sense."

"How *should* she have?" Sir Claude tried again to shine a little at the companion of his walk. "How at any rate is it proved by her going out with me?"

"Don't ask *me*; ask that woman. She drivels when she does n't rage," Mrs. Beale declared.

"And she leaves the child?"

"She leaves the child," said Mrs. Beale with great emphasis and looking more than ever over Maisie's head.

In this position suddenly a change came into her face, caused, as the others could the next thing see, by the re-appearance of Mrs. Wix in the doorway which, on coming in at Sir Claude's heels, Maisie had left gaping. "I *don't* leave the child—I don't, I don't!" she thundered from the threshold, advancing upon the opposed three but addressing herself directly to Maisie. She was girded—positively harnessed—for departure, arrayed as she had been arrayed on her advent and armed with a small fat rusty reticule which, almost in the manner of a battle-axe, she brandished in support of her words. She had clearly come straight from her room, where Maisie in an instant guessed she had directed the removal of her minor effects. "I don't leave you till I've given you another chance. Will you come *with* me?"

Maisie turned to Sir Claude, who struck her as having been removed to a distance of about a mile. To Mrs. Beale she turned no more than Mrs. Beale had turned: she felt as if already their difference had been disclosed. What had come out about that in the scene between the two women? Enough came out now, at all events, as she put it practically to her stepfather. "Will *you* come? Won't you?" she enquired as if she had not already seen that she should have to give him up. It was the last flare of her dream. By this time she was afraid of nothing.

"I should think you'd be too proud to ask!" Mrs. Wix interposed. Mrs. Wix was herself conspicuously too proud.

But at the child's words Mrs. Beale had fairly bounded. "Come away from *me*, Maisie?" It was a wail of dismay and reproach, in which her stepdaughter was astonished to read that she had had no hostile consciousness and that if she had been so actively grand it was not from suspicion, but from strange entanglements of modesty.

Sir Claude presented to Mrs. Beale an expression positively sick. "Don't put it to her *that* way!" There had indeed been something in Mrs. Beale's tone, and for a moment our young lady was reminded of the old days in which so many of her friends had been "compromised."

This friend blushed; she was before Mrs. Wix, and though she bridled she took the hint. "No—it is n't the way." Then she showed she knew the way. "Don't be a still bigger fool, dear, but go straight to your room and wait there till I can come to you."

Maisie made no motion to obey, but Mrs. Wix raised a hand that forestalled every evasion. "Don't move till you've heard me. *I'm* going, but I must first understand. Have you lost it again?"

Maisie surveyed—for the idea of a describable loss—the immensity of space. Then she replied lamely enough: "I feel as if I had lost everything."

Mrs. Wix looked dark. "Do you mean to say you *have* lost what we found together with so much difficulty two days ago?" As her pupil failed of response she continued: "Do you mean to say you've already forgotten what we found together?"

Maisie dimly remembered. "My moral sense?"

"Your moral sense. *Have n't* I, after all, brought it out?" She spoke as she had never spoken even in the schoolroom and with the book in her hand.

It brought back to the child's recollection how she sometimes could n't repeat on Friday the sentence that had been glib on Wednesday, and she dealt all feebly and ruefully with the present tough passage. Sir Claude and Mrs. Beale stood there like visitors at an "exam." She had indeed an instant a whiff of the faint flower that Mrs. Wix pretended to have plucked and now with such a peremptory hand thrust at her nose. Then it left her, and, as if she were sinking with a slip from a foothold, her arms made a short jerk. What this jerk represented was the spasm within her of something still deeper than a moral sense. She looked at her examiner; she looked at the visitors; she felt the rising of the tears she had kept down at the station. They had nothing—no, distinctly nothing—to do with her moral sense. The only thing was the old flat shameful schoolroom plea. "I don't know—I don't know."

"Then you've lost it." Mrs. Wix seemed to close the book as she fixed the straighteners on Sir Claude. "You've nipped it in the bud. You've killed it when it had begun to live."

She was a newer Mrs. Wix than ever, a Mrs. Wix high and great; but Sir Claude was not after all to be treated as a little boy with a missed lesson. "I've not killed anything," he said; "on the contrary I think I've produced life. I don't know what to call it—I have n't even known how decently to deal with it, to approach it; but, whatever it is, it's the most beautiful thing I've ever met—it's exquisite, it's sacred." He had his hands in his pockets and, though a trace of the sickness he had just shown perhaps lingered there, his face bent itself with extraordinary gentleness on both the friends he was about to lose. "Do you know what I came back for?" he asked of the elder.

"I think I do!" cried Mrs. Wix, surprisingly unmollified and with the heat of her late engagement with Mrs. Beale still on her brow. That lady, as if a little besprinkled by such turns of the tide, uttered a loud inarticulate protest and, averting herself, stood a moment at the window.

"I came back with a proposal," said Sir Claude.

"To me?" Mrs. Wix asked.

"To Maisie. That she should give you up."

"And does she?"

Sir Claude wavered. "Tell her!" he then exclaimed to the child, also turning away as if to give her the chance. But Mrs. Wix and her pupil stood confronted in silence, Maisie whiter than ever—more awkward, more rigid and yet more dumb. They looked at each other hard, and as nothing came from them Sir Claude faced about again. "You won't tell her?—you can't?" Still she said nothing; whereupon, addressing Mrs. Wix, he broke into a kind of ecstasy. "She refused—she refused!"

Maisie, at this, found her voice. "I did n't refuse. I did n't," she repeated.

It brought Mrs. Beale straight back to her. "You accepted, angel—you accepted!" She threw herself upon the child and, before Maisie could resist, had sunk with her upon the sofa, possessed of her, encircling her. "You've given her up already, you've given her up for ever, and you're ours and ours only now, and the sooner she's off the better!"

Maisie had shut her eyes, but at a word of Sir Claude's they opened. "Let her go!" he said to Mrs. Beale.

"Never, never, never!" cried Mrs. Beale. Maisie felt herself more compressed.

"Let her go!" Sir Claude more intensely repeated. He was looking at Mrs. Beale and there was something in his voice. Maisie knew from a loosening of arms that she had become conscious of what it was; she slowly rose from the sofa, and the child stood there again dropped and divided. "You're free—you're free," Sir Claude went on; at which

Maisie's back became aware of a push that vented resentment and that placed her again in the centre of the room, the cynosure of every eye and not knowing which way to turn.

She turned with an effort to Mrs. Wix. "I did n't refuse to give you up. I said I would if *he'd* give up—!"

"Give up Mrs. Beale?" burst from Mrs. Wix.

"Give up Mrs. Beale. What do you call that but exquisite?" Sir Claude demanded of all of them, the lady mentioned included; speaking with a relish as intense now as if some lovely work of art or of nature had suddenly been set down among them. He was rapidly recovering himself on this basis of fine appreciation. "She made her condition—with such a sense of what it should be! She made the only right one."

"The only right one?"—Mrs. Beale returned to the charge. She had taken a moment before a snub from him, but she was not to be snubbed on this. "How can you talk such rubbish and how can you back her up in such impertinence? What in the world have you done to her to make her think of such stuff?" She stood there in righteous wrath; she flashed her eyes round the circle. Maisie took them full in her own, knowing that here at last was the moment she had had most to reckon with. But as regards her stepdaughter Mrs. Beale subdued herself to a question deeply mild. "*Have* you made, my own love, any such condition as that?"

Somehow, now that it was there, the great moment was not so bad. What helped the child was that she knew what she wanted. All her learning and learning had made her at last learn that; so that if she waited an instant to reply it was only from the desire to be nice. Bewilderment had simply gone or at any rate was going fast. Finally she answered. "Will you give *him* up? Will you?"

"Ah leave her alone—leave her, leave her!" Sir Claude in sudden supplication murmured to Mrs. Beale.

Mrs. Wix at the same instant found another apostrophe. "Is n't it enough for you, madam, to have brought her to discussing your relations?"

Mrs. Beale left Sir Claude unheeded, but Mrs. Wix could make her flame. "My relations? What do you know, you hideous creature, about my relations, and what business on earth have you to speak of them? Leave the room this instant, you horrible old woman!"

"I think you had better go—you must really catch your boat," Sir Claude said distressfully to Mrs. Wix. He was out of it now, or wanted to be; he knew the worst and had accepted it: what now concerned him was to prevent, to dissipate vulgarities. "Won't you go—won't you just get off quickly?"

"With the child as quickly as you like. Not without her." Mrs. Wix was adamant.

"Then why did you lie to me, you fiend?" Mrs. Beale almost yelled. "Why did you tell me an hour ago that you had given her up?"

"Because I despaired of her—because I thought she had left me." Mrs. Wix turned to Maisie. "You were *with* them—in their connexion. But now your eyes are open, and I take you!"

"No you don't!" and Mrs. Beale made, with a great fierce jump, a wild snatch at her stepdaughter. She caught her by the arm and, completing an instinctive movement, whirled her round in a further leap to the door, which had been closed by Sir Claude the instant their voices had risen. She fell back against it and, even while denouncing and waving off Mrs. Wix, kept it closed in an incoherence of passion. "You don't take her, but you bundle yourself: she stays with her own people and she's rid of you! I never heard anything so monstrous!" Sir Claude had rescued Maisie and kept hold of her; he held her in front of him, resting his hands very lightly on her shoulders and facing the loud adversaries. Mrs. Beale's flush had dropped; she had turned pale with a splendid wrath. She kept protesting and dismissing Mrs. Wix; she glued her back to the door to prevent Maisie's flight; she drove out Mrs. Wix by the window or the chimney. "You're a nice one—'discussing relations'—with your talk of our 'connexion' and your insults! What in the world's our connexion but the love of the child who's our duty and our life and who holds us together as closely as she originally brought us?"

"I know, I know!" Maisie said with a burst of eagerness. "I did bring you."

The strangest of laughs escaped from Sir Claude. "You did bring us—you did!" His hands went up and down gently on her shoulders.

Mrs. Wix so dominated the situation that she had something sharp for every one. "There you have it, you see!" she pregnantly remarked to her pupil.

"*Will* you give him up?" Maisie persisted to Mrs. Beale.

"To *you*, you abominable little horror?" that lady indignantly enquired, "and to this raving old demon who has filled your dreadful little mind with her wickedness? Have you been a hideous little hypocrite all these years that I've slaved to make you love me and deludedly believed you did?"

"I love Sir Claude—I love *him*," Maisie replied with an awkward sense that she appeared to offer it as something that would do as well. Sir Claude had continued to pat her, and it was really an answer to his pats.

"She hates you—she hates you," he observed with the oddest quietness to Mrs. Beale.

His quietness made her blaze. "And you back her up in it and give me up to outrage?"

"No; I only insist that she's free—she's free."

Mrs. Beale stared—Mrs. Beale glared. "Free to starve with this pauper lunatic?"

"I'll do more for her than *you* ever did!" Mrs. Wix retorted. "I'll work my fingers to the bone."

Maisie, with Sir Claude's hands still on her shoulders, felt, just as she felt the fine surrender in them, that over her head he looked in a certain way at Mrs. Wix. "You need n't do that," she heard him say. "She has means."

"Means?—Maisie?" Mrs. Beale shrieked. "Means that her vile father has stolen!"

"I'll get them back—I'll get them back. I'll look into it." He smiled and nodded at Mrs. Wix.

This had a fearful effect on his other friend. "Have n't *I* looked into it, I should like to know, and have n't I found an abyss? It's too inconceivable—your cruelty to me!" she wildly broke out. She had hot tears in her eyes.

He spoke to her very kindly, almost coaxingly. "We'll look into it again; we'll look into it together. It *is* an abyss, but he *can* be made—or Ida can. Think of the money they're getting now!" he laughed. "It's all right, it's all right," he continued. "It would n't do—it would n't do. We *can't* work her in. It's perfectly true—she's unique. We're not good enough—oh no!" and, quite exuberantly, he laughed again.

"Not good enough, and that beast *is*?" Mrs. Beale shouted.

At this for a moment there was a hush in the room, and in the midst of it Sir Claude replied to the question by moving with Maisie to Mrs. Wix. The next thing the child knew she was at that lady's side with an arm firmly grasped. Mrs. Beale still guarded the door. "Let them pass," said Sir Claude at last.

She remained there, however; Maisie saw the pair look at each other. Then she saw Mrs. Beale turn to her. "I'm your mother now, Maisie. And he's your father."

"That's just where it is!" sighed Mrs. Wix with an effect of irony positively detached and philosophic.

Mrs. Beale continued to address her young friend, and her effort to be reasonable and tender was in its way remarkable. "We're representative, you know, of Mr. Farange and his former wife. This person represents mere illiterate presumption. We take our stand on the law."

"Oh the law, the law!" Mrs. Wix superbly jeered. "You had better indeed let the law have a look at you!"

"Let them pass—let them pass!" Sir Claude pressed his friend hard—he pleaded.

But she fastened herself still to Maisie. "*Do* you hate me, dearest?"

Maisie looked at her with new eyes, but answered as she had answered before. "Will you give him up?"

Mrs. Beale's rejoinder hung fire, but when it came it was noble. "You should n't talk to me of such things!" She was shocked, she was scandalised to tears.

For Mrs. Wix, however, it was her discrimination that

266

was indelicate. "You ought to be ashamed of yourself!" she roundly cried.

Sir Claude made a supreme appeal. "Will you be so good as to allow these horrors to terminate?"

Mrs. Beale fixed her eyes on him, and again Maisie watched them. "You should do him justice," Mrs. Wix went on to Mrs. Beale. "We've always been devoted to him, Maisie and I—and he has shown how much he likes us. He would like to please her; he would like even, I think, to please me. But he hasn't given you up."

They stood confronted, the step-parents, still under Maisie's observation. That observation had never sunk so deep as at this particular moment. "Yes, my dear, I haven't given you up," Sir Claude said to Mrs. Beale at last, "and if you'd like me to treat our friends here as solemn witnesses I don't mind giving you my word for it that I never never will. There!" he dauntlessly exclaimed.

"He can't!" Mrs. Wix tragically commented.

Mrs. Beale, erect and alive in her defeat, jerked her handsome face about. "He can't!" she literally mocked.

"He can't, he can't, he can't!"—Sir Claude's gay emphasis wonderfully carried it off.

Mrs. Beale took it all in, yet she held her ground; on which Maisie addressed Mrs. Wix. "Shan't we lose the boat?"

"Yes, we shall lose the boat," Mrs. Wix remarked to Sir Claude.

Mrs. Beale meanwhile faced full at Maisie. "I don't know what to make of you!" she launched.

"Good-bye," said Maisie to Sir Claude.

"Good-bye, Maisie," Sir Claude answered.

Mrs. Beale came away from the door. "Good-bye!" she hurled at Maisie; then passed straight across the room and disappeared in the adjoining one.

Sir Claude had reached the other door and opened it. Mrs. Wix was already out. On the threshold Maisie paused; she put out her hand to her stepfather. He took it and held it a moment, and their eyes met as the eyes of those who have done for each other what they can. "Good-bye," he repeated.

"Good-bye." And Maisie followed Mrs. Wix.

They caught the steamer, which was just putting off, and, hustled across the gulf, found themselves on the deck so breathless and so scared that they gave up half the voyage to letting their emotion sink. It sank slowly and imperfectly; but at last, in mid-channel, surrounded by the quiet sea, Mrs. Wix had courage to revert. "I did n't look back, did you?"

"Yes. He was n't there," said Maisie.

"Not on the balcony?"

Maisie waited a moment; then "He was n't there" she simply said again.

Mrs. Wix also was silent a while. "He went to *her*," she finally observed.

"Oh I know!" the child replied.

Mrs. Wix gave a sidelong look. She still had room for wonder at what Maisie knew.

Page	1st English edition	New York edition
131	Mrs. Beale was now too intent in seeing what became of the others.	Mrs. Beale was now, in her instant vigilance, too immensely "on"
135	There was an unaccustomed geniality in his enjoyment of her wonder	He seemed to treat her wonder as a positive tribute
	his unexpected gentleness was too mystifying.	yet the way he spared them made her rather uneasy too.
138	as if he had drawn—rather red with the confusion of it—the pledge of her preparation from her tears	as if, though he was so stupid all through, he had let the friendly suffusion of her eyes yet tell him she was ready for anything
143	made the word boom out	made the ugly word—ugly enough at best—sound flat and low
147	the pride momentarily suggested by Beale's association with so much taste.	the pleasure briefly taken in Beale's command of such elegance
	There was no taste in his association with the	There was no command of elegance in his having exposed her to the approach of the
148	being found in that punishable little attitude toward	being so consciously and gawkily below
150	her property under her pillow.	her property gathered into a knotted handkerchief, the largest that could be produced and lodged under her pillow.
151	These days would be exciting indeed	These days would become terrific like the Revolutions she had learnt by heart in Histories
179	of a redness really so vivid as to be feverish	of a redness associated in Maisie's mind at *that* pitch either with measles or with "habits"
182	a perceptive person	a young person with a sharpened sense for latent meanings
184	". . . I almost admire her!" she proclaimed.	". . . I almost admire her!" she quavered
223	a mere trousered thing	a mere rough trousered thing, destined at the best probably to grow up a scamp

Page	1st English edition	New York edition
228	with peculiar consideration	as if I wasn't quite a worm
	Her pupil faltered a few seconds	Her pupil, though wholly without aspirations to pertness, barely faltered
230	that of such preliminaries ablutions	that of such preliminaries those embodied in a thorough use of soap
261	the tears she had kept down at the station	the tears she had kept down at the station. They had nothing —no, distinctly nothing—to do with her moral sense